TV Video Systems
for the
Hobbyist & Technician

By
L.W. Pena & Brent A. Pena

PROMPT.
PUBLICATIONS

An Imprint of
Howard W. Sams & Company
Indianapolis, Indiana

REVISED FIRST EDITION, 1996

PROMPT® Publications is an imprint of Howard W. Sams & Company, a Bell Atlantic Company, 2647 Waterfront Parkway, E. Dr., Suite 300, Indianapolis, IN 46214-2041.

This book was originally developed and published as *Installing TV Video Systems* by Master Publishing, Inc., 522 Cap Rock Drive, Richardson, Texas 75080-2036.

International Standard Book Number: 0-7906-1082-5

Editors: Bob Hurst, Gerald Luecke, David A. Wolf
Text Design and Artwork: Plunk Design, Dallas, TX
Cover Design by: Suzanne Lincoln
Photographs: All photographs that do not have a source identification are either courtesy of Radio Shack, the authors, or Master Publishing, Inc.

Trademark Acknowledgments:
- Radio Shack® is a registered trademark of Tandy Corporation.
- RCA® is a registered trademark of Radio Corporation of America.

All terms in this book that are known or suspected to be trademarks or service marks have been appropriately capitalized or marked with trademarks or service mark symbols at the first mention of the trademark or service mark. Every effort has been made to supply complete and accurate information. PROMPT® Publications, Howard W. Sams & Company, Bell Atlantic, and Master Publishing, Inc., cannot attest to the accuracy of this information. Use of a term in this book should not be regarded as affecting the validity of any trademark or service mark.

Printed in the United States of America

9 8 7 6 5 4 3 2 1

Table of Contents

Preface

The purpose of *TV Video Systems for the Hobbyist & Technician* is to explain the various options that are available for receiving TV pictures and sound. It explains hard-line cable systems, home satellite systems, wireless cable TV systems, and the new small-dish digital systems, DSS®. It shows how the systems are installed, explains the advantages and disadvantages, shows installations that can be done by the "do-it-yourselfer," deals with typical problems and how to detect and repair them, and shows multiple ways of interconnecting available TV signal sources into an integrated system.

The first two chapters of *TV Video Systems for the Hobbyist & Technician* provide an overview of the systems and their relative costs within four major regions of the country. There is a chapter that details the system components, followed by two chapters on how TV video systems are installed in apartments and in homes. Common connectors are identified, how to add an A/B switch and a signal splitter, and the interconnection of an "off-the-air" antenna into a DSS system are explained.

Chapter 7, which is the main chapters for readers interested in doing their own installations, provides discussion of equipment, source of supplies, coaxial cable preparation and F-connector attachment, before it shows step-by-step instructions on specific installations. Below the house and attic installations are covered in detail. Prior to Chapter 7, there is a chapter that discusses typical causes of poor TV picture quality.

The book concludes with a chapter on how to add accessories with many detailed diagrams showing multiple interconnections, and a chapter on troubleshooting and repairing systems. Charts of typical problems and their solutions are included.

TV Video Systems for the Hobbyist & Technician was written to inform the reader about the choices available to receive TV signals. It is also for the technician, hobbyists, technically-interested consumer, and "do-it-yourselfer" who would like to delve into how things work and how things are put together. That was our goal; we hope we have succeeded.

LP, BP, MPI

Chapter 1
The Different TV Video Systems

INTRODUCTION TO TV VIDEO SYSTEMS

TV video systems, which include cable TV and home satellite systems, are some of the most popular forms of entertainment in America today. Consumers are no longer limited to just a handful of TV channels. A multi-billion-dollar industry has evolved to provide viewers with a plethora of entertainment choices and a variety of ways to obtain them!

The Four Different Services

There are four primary methods of delivering signals to subscribers.

1. The first is hardline cable, which is the traditional cable TV service most consumers have. This system is composed of a large network of cable lines which run throughout a city, into neighborhoods, then to individual homes and apartments.
2. The second method is the home satellite system, which requires having a large satellite dish on the homeowner's premises. This satellite dish is used to receive downlink transmissions from orbiting satellites. The dish is steerable, so that it may receive signals from many satellite systems, both foreign and domestic.
3. The third method of getting a signal to the subscriber is so-called "wireless cable TV." This service delivers programs by microwaves instead of a large network of cables. It uses a small receiving antenna, installed on the subscriber's home.
4. The fourth method of delivery is the Digital Satellite System® (DSS)®. With DSS, the subscriber forgoes the large satellite dish for a smaller stationary receiving dish. Because it is stationary, only one satellite may be received. The subscriber obtains a 'package' of programming from one service provider. The DSS may be purchased outright and programming negotiated for separately, or leased together with a programming package.

As a consumer, having so many options to choose from can be fun, but a little confusing. Should you buy a large satellite home system or is the traditional hardline cable TV better for you? Is investing in a small Digital Satellite System right for you or will wireless cable provide all the programs you want, for a lower investment? These are just some of the questions you need to ask yourself before making a final decision. Where can you turn to get the facts necessary to make a knowledgeable decision?

ABOUT THIS BOOK

It is the purpose of this book to describe the various TV video services available to consumers. It will show the different components of the four primary systems, and how they operate. The book will also show you some of the common installations and how they may be easily performed.

Home and Apartments

New home installations and adding additional outlets are among the many subjects covered. Along with the typical cable TV installations, you'll discover how to correctly add accessories. Interconnecting VCRs, closed-captioned devices and stereos will be fully explained and illustrated. If you are concerned about making a mistake, don't worry! An extensive troubleshooting section is included so that you may check your work.

Not in a Suburban Neighborhood?

What about rural and mountain residents who live in an area where cable TV does not exist? Good news! Three of the four home video reception systems are available for people in these areas. This book will show the benefits and drawbacks of having any of the three systems. It also includes what is needed, the limitations, and the expenses that can be incurred.

If you live in an apartment, this book is for you, too. Along with general apartment antenna systems, there is a section on adding outlets.

When you are ready to make your final decision about a TV video system, you will do so with confidence. Knowing the facts about what the different systems have to offer will allow you to make a sound decision without falling victim to a high-pressure sales pitch. Being an educated consumer means saving time, money, and energy.

HARDLINE TV VIDEO SYSTEM

What is a hardline TV video system? Hardline TV video service is provided by your local cable TV company. Most often it is referred to as cable TV or just "cable." The average cable TV company provides several layers or 'tiers' of programming. Basic service usually consists of your local broadcast TV channels plus several channels with 24-hour news, sports, weather and some movies. For an additional charge, these companies will also provide you with premium channels such as HBO, Cinemax, and Showtime. If you prefer first-run movies or live special events (primarily sports and music), pay-per-view channels can also be found on today's cable systems. Many cable TV companies also provide stereo music service (up to 100 channels!), commercial-free.

Signals from the different cable networks are transmitted ("uplinked") to satellites in geostationary orbit above the Earth. See *Figure 1-1*. These signals are then retransmitted by the satellites and received ("downlinked") by cable TV companies around the country. Your local cable TV company combines the satellite signals with your local broadcast UHF and VHF TV stations. This combined signal is amplified and sent throughout the local cable company's network by way of semi-rigid cables called feederlines. See *Figure 1-2*. Feederline is the technical term

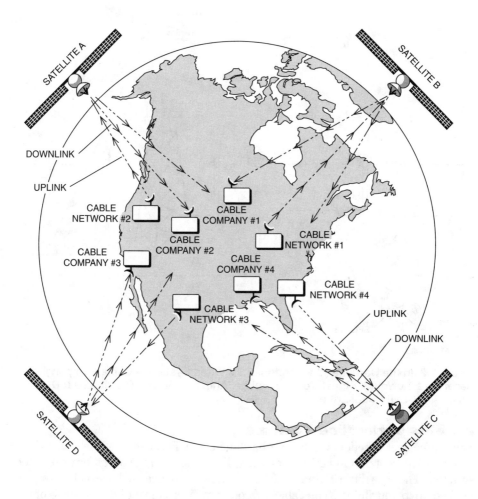

Figure 1-1. Cable Networks Transmit Signals to Orbiting Satellites on Uplinks. Cable Companies Receive These Signals on Downlinks, Then Distribute Them to Subscribers.

used to describe the main distribution cable installed on telephone poles and underground throughout a cable TV company's service area. Amplifiers are installed along the feederlines to maintain signal quality across the entire network. The cable signal is transmitted through a "tap," which is connected along a feederline. To receive the service, flexible coaxial cable is connected to the tap and is run to a groundblock attached at an entrance point in your home. From the groundblock the cable enters your home and goes to a device called a "splitter." The type of splitter you will need is determined by the number of TVs (and FM stereos) in your home. For example, you will need a two-way splitter if you have two TVs in your home, or a three-way splitter if you have two TVs and a stereo.

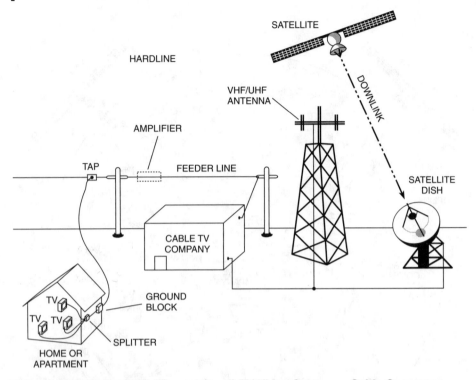

Figure 1-2. In the Hardline (Conventional) TV Video System, a Cable Company Receives a Downlink Signal from a Satellite, Amplifies It, and Distributes It Over Feederlines to Its Subscribers.

Benefits of Hardline Systems

There are several benefits in having hardline cable TV instead of the other video services. The cost of regular hardline cable TV is less than the other forms of video systems. There is little or no initial subscriber expense to have cable service hooked up to your home. The cable TV company bears most of the expense of the hardware and labor. Basic programming service charges are usually a nominal monthly fee. Where subscribers incur large charges is in the purchase of multiple premium channels. Depending on the number of premium channels you select, your billing could range from $15 to $100+ each month.

A good reason for choosing hardline cable TV is that most urbanized areas of the country are already wired up. The actual hookup to your home may take less than an hour. When properly installed, hardline cable TV is the least noticeable service of all the video systems. Cable lines can be run underground or high enough overhead so that no one will notice the connection. The only visible sign may be at the entry of the coaxial cable into the attic or basement.

Another popular reason to have hardline cable TV is the speed of installation. Since your neighborhood or even your home may be pre-wired for cable, you can order cable service and be enjoying your expanded TV programming within a few days! As stated above, typical installation time is less than an hour. This means less time waiting for service, once the installer arrives. Most do-it-yourselfers can easily master the skills necessary to expand cable hookups within their homes. Most tools and equipment you need may be purchased at a local electronics store.

Installing Your Own Outlets

In most cases, a do-it-yourself home project that adds extra outlets is legal. What is considered illegal is connecting any kind of cable from the tap (at the feederline) to your house for free use of service. This is considered piracy or theft from the local cable TV company.

Also be aware of some of the new cable TV laws from the 1992 Cable TV Act. The new law prevents the cable TV company from charging you for additional outlets. However, they can charge you rent on any equipment such as converter boxes and remote controls. This is how extra charges get built up on your account. To discover the actual prices of these and other charges, contact your local cable TV operator. Their telephone number is listed in your local telephone directory.

BIG-DISH SATELLITE TV VIDEO SYSTEMS

Satellite TV systems are well-established methods to receive news, weather, sports, and movies. Before most areas were wired up with cable TV by the late 1980's, many viewers used home satellite dishes to receive entertainment and information. A typical system is shown in *Figure 1-3*. Satellite dish owners were not limited to North American programming. Even in the late 1970's, with several satellites to choose from, dish owners were able to view international news and entertainment.

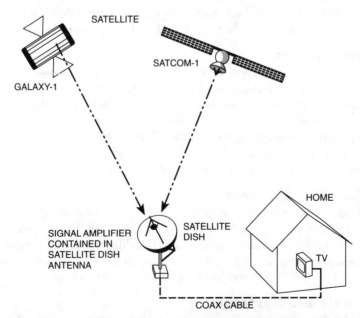

Figure 1-3. Satellite TV Video Systems Consist of a Satellite Dish Antenna That Receives a Downlink Signal from Various Satellites, Amplifies It, and Distributes It Over Coaxial Cable to the Subscriber's TV Set.

A window to the world was made available to even the most remote rural areas all over America. And early on, programming wasn't scrambled!

Today, are there as many as 250 channels available for viewing on home satellite systems? How did this variety come about? The answers lie in the method of delivery of the signals. Television networks and cable TV programming providers throughout the world distribute all of their programming to their affiliates by way of satellite transmissions. Before scrambling became almost universal, home satellite enthusiasts were able to receive the same programming intended for the network affiliates and cable TV companies free of charge. Today, there are still satellite networks that beam signals specifically for free home satellite viewing, but most large-dish satellite downlink signals are beamed to affiliates using a special de-scrambler. The de-scrambler is usually built into the satellite receiver.

Purchasing a Satellite System

If thinking about purchasing a home satellite system, take time to survey your home. Remember, the satellite dish needs an unobstructed view to the south. The signal will *not* pass through trees. If you have a many high trees in your yard it could mean you're out of luck with a ground-mounted dish.

It is possible to mount the satellite dish on top of a tall pole. A 20- to 30-foot pole can be attached to your house, which would put the satellite dish above the roof of your home. The advantage of having a tall, mounted dish, is in it not taking up your backyard area. It also means your satellite dish will be out of reach of unauthorized or young hands.

Before acting, you should also check your homeowner's deed and local zoning ordinances to make sure that a large dish is allowable.

Available Systems

Let us look at the different types of satellite systems that are available. The most common sizes are seven feet and ten feet in diameter. The size of the satellite dish you purchase is dependent on the amount of room you have available for a home satellite system. The larger the dish, the better the reception. This is not to say that a seven-foot dish does not get a good picture. It does. However, a ten-foot dish will deliver a better quality picture. Remember, a bigger dish gathers more signal, which produces a higher resolution picture on your TV set.

Basic Satellite Systems

The most basic home satellite systems are designed to remain stationary, and deluxe models can pick up multiple satellite signals. The basic model faces one direction and is pointed at only one satellite. This kind of system depends on one satellite, such as Galaxy 1, for its entire programming. This type of home satellite system is the least expensive and has very basic programming to offer.

Deluxe Satellite Systems

You will find deluxe home satellite systems that will pick up over 250 channels and multiple satellites. Pay-per-view, HBO®, Cinemax®, Showtime®, The Movie Channel®, many sports channels, and 24-hour news and weather from around the world are available with a deluxe system. You can buy systems with wireless remote controls to make channel surfing easy. Stereo sound can be connected at anytime to enhance the audio quality of what you are viewing. When shopping for a home satellite system, know what you want before talking with a salesman. To ensure you are getting a fair price, talk to at least three different dealers before making a purchase decision.

A Satellite System — Plus and Minus

There are many benefits to having a home satellite system. Rural and mountain residents can have access to information and entertainment once only available to traditional cable subscribers. Home satellite systems offer the broadest range of video and audio entertainment. With over 250 channels to choose from, variety is never a problem. There are some drawbacks to the home satellite system. First is the large investment in the system. The price for these systems range from a few hundred to several thousand dollars. But the system is your, forever. Another drawback is the lack of local stations. To receive local channels, you will need a conventional off-the-air antenna. As with any home video system, you must decide if the benefits outweigh the tradeoffs. Luckily you have some help: *TV Video Systems for the Hobbyist & Technician*!

WIRELESS CABLE TV VIDEO SYSTEM

What is "wireless" cable TV? Wireless cable TV is a new way to bring cable TV-type programs into your home. The technical term for wireless cable TV is Multichannel Multipoint Distribution Service (MMDS). It operates in the 2500 - 2686 MHz (2.5 - 2.686 GHz) microwave band and has 28 channels.

This is how it works: The signal is broadcast from a high structure in your area. Wireless cable companies broadcast their signals to subscribers in a 30-mile radius, usually from a downtown site. It is then received by a microwave antenna attached to your house. See *Figure 1-4.* If your home is pre-wired for cable TV as shown in *Figure 1-5,* or you had cable installed in the past, your existing coaxial cable may be usable. Ask the installer/representative from the company if the existing cable is compatible.

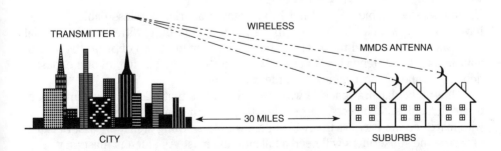

Figure 1-4. 28 Channels of Programming are Sent by Microwave from a Downtown Site to Subscribers' Homes in Suburbs. Typical Range is Less Than 30 Miles.

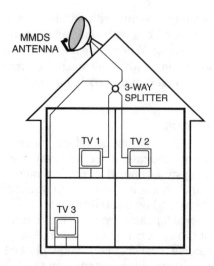

Figure 1-5. MMDS Antenna is Mounted on Rooftop Mast (up to 35'). Signal May Be Distributed Over Pre-Existing Cable Wiring in the Home.

Like other TV video systems, wireless cable TV offers the subscriber basic service. Channels such as CNN®, The Weather Channel®, VH-1®, and MTV® make up basic service programming. Also, like traditional cable TV, premium channels and pay-per-view channels can be added to the basic programming for an additional charge.

The Benefits of Wireless Cable

There are several benefits to having wireless cable TV. If you live in an area that is newly developed, or where there are no existing cable lines, wireless cable may be a good choice. The installation of the system generally takes longer than it does for conventional cable. However, it is less time-consuming than waiting for the cable company to install new cable lines in your area! In most cases, installation takes a lot less time than delivery and installation of a full-size satellite dish or Digital Satellite System.

In some areas of the country, many wired-cable TV subscribers are switching to wireless cable TV. Why? Some are switching as a form of protest against their local cable system. Subscribers cite cost as one major reason for changing. While the installation for wireless cable TV is more expensive than conventional cable, the monthly basic service charge is about $10 less than conventional cable. The tradeoff is that the number of wireless channels available is limited to a maximum of 28 (instead of 50 or 100). Subscribers also list service as a reason for their conversion. Because the customer base is smaller than the wired-cable service, there are more technicians and installers available to serve subscribers.

Wireless Cable Isn't Perfect

Having wireless cable TV will not bring total euphoria. Wireless cable TV does have a few drawbacks. The subscriber's home has to be outfitted with a special microwave antenna and mast. The antenna is mounted on top of or on the side of your house. The antenna is about the size of a newspaper. The height of the mast is dependent upon your distance from the transmitter.

One particular drawback of wireless cable TV is that microwaves cannot penetrate through solid objects. Large office buildings, mountain ranges, and tall trees with leaves create barriers that will not allow even a strong signal to pass. Because of this, some subscribers will need a tall antenna mast. An antenna mast may be as high as 35' to clear signal obstructions. This may seem somewhat extreme, but remember that the objective is to provide the subscriber with a clear signal.

Even with its drawbacks, wireless cable TV is a good buy. The subscriber gets a clear picture, good customer service, and a slightly lower monthly bill than conventional cable TV.

DIGITAL SATELLITE SYSTEM TV

Direct Broadcast Satellite® (DBS)® or Digital Satellite System (DSS), is the latest technological advancement in home TV video systems. Digital Satellite System delivers its signal through direct satellite-to-earth transmission, instead of a coaxial cable or MMDS microwave link. A major departure from all previous forms of television is the digital, rather than analog, mode of the signal. A digital signal means laser disc audio and video quality with no signal fading due to weather conditions. A digital signal is either there or it isn't; there isn't any deterioration or distortion from weather. Like all other satellite distribution methods, signals are uplinked from a ground station to a satellite 'parked' in geostationary orbit, then downlinked to the subscriber's dish. See *Figure 1-6.*

Little Dish, Big Advantage

At the receiving end of this signal is a small satellite dish. The dish is about 18 inches wide and is affixed to the subscriber's home. The size of the dish is an advantage of this video system. Because it is only 18 inches in diameter, it is unobtrusive. The subscriber has the benefit of a home satellite system without the large satellite dish. It is easily installed where deed restrictions and municipal ordinances restrict the use of larger dishes or outside antennas.

Almost any home can have a Digital Satellite System dish. The only restriction is having an unobstructed southern view of the horizon, just as with the larger satellite dishes. Why? The orbits of geostationary satellites require that they be over the equator, which is due south from all points in North America.

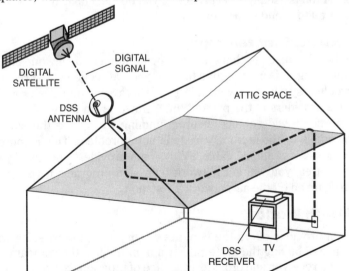

Figure 1-6. Digital Satellite System Transmits 150 Channels of Programs to Small Dish Mounted on Your Home.

Great Pics If You're In the Sticks

Rural and mountain residents will like this new system. The programming for the Digital Satellite System offers a lot of variety, just like traditional cable. It offers the subscriber news, weather, sports, premium and pay-per-view channels. These features may not sound unusually great, but consider the residents' situation. Rural and mountain families have been denied access to traditional cable service because of their location. Either the terrain was too harsh for cable to be laid, or the population was so sparse that these communities were not seen as being economically feasible for cable service to be profitable. Direct broadcast satellite companies have stepped in to fill this vacuum. Digital Satellite System programming offers 150 channels to choose from! News, weather, sports, and movies make up part of the service. Digital Satellite System customers also have crystal-clear audio channels. There are several channels that contain only audio and no commercials. These channels play Classical, Country & Western, Top 40, Rap, R&B, Rock & Roll (Hard, Classic, Soft). There is a channel for almost every musical taste.

DSS Has Some Drawbacks

There are several drawbacks if you purchase your Digital Satellite System TV. There is a large investment in the dish and converter box that comprise the system. The retail price of this dish and receiver is about $700. The price is dependent upon location and availability. Availability itself can be a problem. In some parts of the country, there are waiting lists just to purchase the units!

Once you get your DSS unit, you now face another hurdle. Installation. You can install the unit yourself. Although it is time-consuming and tedious work, some owners do successfully install the units themselves. The average time it takes a professional installer to set up a unit is two-and-a-half to three hours. The typical cost of installation is usually $150 to $200. If you do elect to do it yourself, be patient and be prepared to spend some time on it.

A Word About Programming

The only other drawback to this system is being limited to only one satellite for all your programming. It is not possible to change the direction of the receiver to pick up other satellite systems, as with a large dish. Programming is provided by two major service providers at the present time. However, in fairness, Digital Satellite Systems do offer a wide choice of programming. By giving the subscriber 150 different channels to chose from, variety is not a problem. This is more programming than most conventional cable TV systems provide. A bonus if you are in a really remote area: you will get network television programs if you are currently unable to pick up an off-the-air network TV station.

Best of Both Worlds

One program provider offers the benefits of Direct Digital Satellite without the major expense of buying the system. For a monthly fee, the company will install a DSS dish and receiver, and provide a package of programs for you to view. Check with a home entertainment retailer for more details.

Chapter 2
Contracting for TV Video Services

YOU'VE GOT CHOICES

With four major TV video systems available today, the viewer has some real choices. Not just in the variety of channels available, but also price. Basic programming services are typically provided on all four video systems for a relatively low monthly fee. Depending on the subscriber's preference, numerous premium channels and combination packages are available. Do you want HBO, Cinemax, and Showtime, only? Or do you want all *five* HBO channels? The combination of choices is almost limitless, and, of course, the prices vary with the choices you make.

This section will provide a comparison of approximate rates for the four major TV video systems in the various regions of the country. We will review what services are available, and the costs for these services in each region. Your specific rates will vary depending on your service provider, your particular installation and regulations in effect at the time of installation. Note that service rates *do not* include the cost of installation. Due to labor and material cost fluctuations, and the competitive nature of the marketplace, you are encouraged to contact your local service providers for exact rates. Ask about unadvertised special offers that may be available, especially when it comes to installation and 'get-acquainted' deals — you may be pleasantly surprised! And better informed to make an intelligent choice.

For some types of systems the rate tends to stay the same in all regions of the country. A good example of this is the large-dish home satellite system. After your initial investment of purchasing the satellite dish, you have about 250 channels available for viewing at no charge. When the costs for premium channels and programming packages are added on, monthly rates are incurred. It is important to keep this in mind when you are reading rate information.

REGIONAL COSTS FOR VIDEO CABLE SYSTEMS

Northeast Region

The Northeast Region, shown in *Figure 2-1,* consists of the New England and Mid-Atlantic States: Connecticut, Maine, Massachusetts, New Hampshire, New Jersey, New York, Pennsylvania, Rhode Island and Vermont.

Hardline Cable TV System

In the Northeast, the most basic programming service (often referred to as 'tier one') provided by hardline cable TV systems (CATV) includes channels such as CNN, C-Span®, Headline News®, CMT®, MTV, The Weather Channel, and your

Figure 2-1. Northeast Region of U.S.
Source: U.S. Bureau of the Census, Statistical Abstract of the United States: 1994 (114th Ed.)
Washington, DC, 1994

local VHF and UHF stations. The available number of premium and pay-per-view channels will vary throughout the region. The greater the density of population, the more likely a larger amount of premium and pay-per-view channels will be available. For example, New York City, with a greater concentration of people, will have more specialized channels available than does Elmira, New York. Typical rates for hardline cable TV are as follows:

Hardline Cable TV Rates

Service	Monthly Rate
Basic Service	$20.00 - $25.00
Basic Service with Premium Channels	$25.00 - $40.00

Home Satellite System

Due to their large size, the majority of home satellite system owners will be in less-populated areas (rural or spacious suburbs). Home satellite systems allow the owner to receive over 250 different channels from multiple satellites. CNN, C-Span, Headline News, The Weather Channel, and similar channels, can all be received on the home satellite system. If the viewer misses a showing of a network program, he can redirect his satellite dish to a western satellite to pick up the show again. Because of the time zone difference, northeastern residents have the maximum time to see a later showing of their favorite program. Typical rates are as follows:

Home Satellite System Rates

Service	Monthly Rate
Basic Service	no charge
Basic Service with Premium Channels	$7.00 - $15.00

Wireless Cable TV System

Wireless cable TV (MMDS) is available in limited areas in this region. In those areas where wireless cable TV is available, standard 28-channel service is offered. This basic service includes channels such as CNN, C-Span, Headline News, CMT, MTV, and The Weather Channel. Premium channels are available, but far fewer than with hardline cable TV — perhaps as few as three or four. HBO, Cinemax, and the Disney Channel® are the typical premium channels available on wireless cable TV. Multiple HBO and multiple Cinemax channels are rare on this service. Typical rates are as follows:

Wireless Cable TV Rates

Service	Monthly Rate
Basic Service	$15.00 - $20.00
Basic Service with Premium Channels	$25.00 - $30.00

Digital Satellite System

The Digital Satellite System (DSS) is the answer to the variety problem for rural areas of the Northeastern states. Along with receiving networks such as CNN, C-Span, Headline News, CMT, MTV, and The Weather Channel, the subscriber can choose from a selection of compact-disc-quality audio channels. Viewers in the suburban and rural areas of the Northeastern states will really enjoy this system. For the same price as premium cable TV, the subscriber gets more channels and ownership of the equipment. Although DSS provides the customer with 150 channels, he still needs a regular antenna to receive local VHF and UHF channels. Typical rates are as follows:

Digital Satellite System Rates

Service	Monthly Rate
Basic Service	$15.00 - $20.00
Basic Service with Premium Channels	$30.00 - $35.00

South Region

The South Region of the U.S., shown in *Figure 2-2,* consists of the South Atlantic, East South-Central, and West South-Central States: Alabama, Arkansas, Delaware, Florida, Georgia, Kentucky, Louisiana, Maryland, Mississippi, North Carolina, Oklahoma, South Carolina, Tennessee, Texas, Virginia and West Virginia.

Hardline Cable TV System

In the South, the most basic programming service (often referred to as 'tier one') provided by hardline cable TV systems (CATV) includes channels such as CNN, C-Span, Headline News, CMT, MTV, The Weather Channel, and your local VHF and UHF stations. The available number of premium and pay-per-view channels will vary throughout the region. The greater the density of population, the more likely a larger amount of premium and pay-per-view channels will be available. For example, Houston, with a greater concentration of people, will have more services available for its subscribers than will Joshua, Texas. Typical rates are as follows:

Hardline Cable TV Rates

Service	Monthly Rate
Basic Service	$15.00 - $20.00
Basic Service with Premium Channels	$21.00 - $31.00

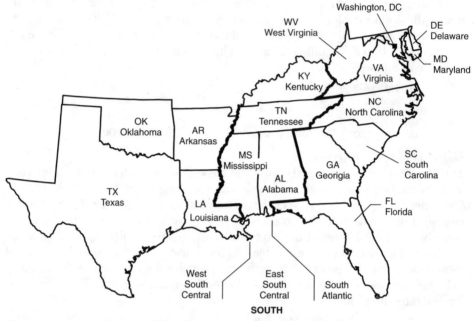

Figure 2-2. South Region of the U.S.
Source: U.S. Bureau of the Census, Statistical Abstract of the United States: 1994 (114th Ed.)
Washington, DC, 1994

Home Satellite System

Due to their large size, the majority of home satellite system owners will be in less-populated areas (rural or spacious suburbs). Home satellite systems allow the owner to receive over 250 different channels, from multiple satellites. CNN, C-Span, Headline News, The Weather Channel, and the like, can all be received with a home satellite system. If the viewer misses a showing of a network program, he can often-times redirect his satellite dish to a Western satellite to pick up the same show at a later time because of multiple transmissions for the different time zones. Typical rates are as follows:

Home Satellite System Rates

Service	Monthly Rate
Basic Service	no charge
Basic Service with Premium Channels	$7.00 - $15.00

Wireless Cable TV System

Wireless cable TV (MMDS) is available in limited areas in this region. In those areas where wireless cable TV is available, standard 28-channel service is offered. This basic service includes channels such as CNN, C-Span, Headline News, CMT,

MTV, and The Weather Channel. Premium channels are available, but far fewer than with hardline cable TV — perhaps as few as three or four. HBO, Cinemax, and the Disney Channel are the typical premium channels available on wireless cable TV. Multiple HBO and multiple Cinemax channels are rare on this service. Typical rates are as follows:

Wireless Cable TV Rates

Service	Monthly Rate
Basic Service	$9.00 - $15.00
Basic Service with Premium Channels	$25.00 - $30.00

Digital Satellite System

The Digital Satellite System (DSS) is the answer to the variety problem for rural areas of the Southern states. Along with receiving networks such as CNN, C-Span, Headline News, CMT, MTV, and The Weather Channel, the subscriber has a selection of compact-disc-quality audio channels, as well. Viewers in the suburban and rural areas of the Southern states will really enjoy this system. For the same price as premium cable TV, the subscriber gets more channels and ownership of the equipment. Although DSS provides the customer with 150 channels, a regular antenna is still needed to receive local VHF and UHF channels. Typical rates are as follows:

Digital Satellite System Rates

Service	Monthly Rate
Basic Service	$15.00 - $20.00
Basic Service with Premium Channels	$30.00 - $35.00

Midwest Region

The Midwest Region, shown in *Figure 2-3,* includes the East North-Central and West North-Central States: Iowa, Illinois, Indiana, Kansas, Minnesota, Michigan, Missouri, North Dakota, Nebraska, Ohio, South Dakota and Wisconsin.

Hardline Cable TV System

In the Midwest, the most basic programming service (often referred to as "tier one") provided by hardline cable TV systems (CATV) includes channels such as CNN, C-Span, Headline News, CMT, MTV, The Weather Channel, and your local VHF and UHF stations. The available number of premium and pay-per-view channels will vary throughout the region. The greater the density of population, the more likely a larger amount of premium and pay-per-view channels will be available. For example, Chicago, with a greater concentration of people, will have more specialized channels available than will Hudson, Wisconsin. Typical rates are as follows:

Hardline Cable TV Rates

Service	Monthly Rate
Basic Service	$15.00 - $20.00
Basic Service with Premium Channels	$25.00 - $30.00

Figure 2-3. Midwest Region of the U.S.
Source: U.S. Bureau of the Census, Statistical Abstract of the United States: 1994 (114th Ed.) Washington, DC, 1994

Home Satellite System

Due to their large size, the majority of home satellite system owners will be in less-populated areas (rural or spacious suburbs). Home satellite systems allow the owner to receive over 250 different channels, from multiple satellites. CNN, C-Span, Headline News, The Weather Channel, and the like, can all be received with a home satellite system. Typical rates are as follows:

Home Satellite System Rates

Service	Monthly Rate
Basic Service	no charge
Basic Service with Premium Channels	$7.00 - $15.00

Wireless Cable TV System

Wireless cable TV (MMDS) is available in limited areas of this region. In those areas where wireless cable TV is available, standard 28-channel service is offered. This basic service includes channels such as CNN, C-Span, Headline News, CMT, MTV, and The Weather Channel. Premium channels are available, but far fewer than with hardline cable TV — perhaps as few as three or four. HBO, Cinemax, and the Disney Channel are the typical premium channels available on wireless cable TV. Multiple HBO and multiple Cinemax channels are rare on this service. Typical rates are as follows:

Wireless Cable TV Rates

Service	Monthly Rate
Basic Service	$17.00 - $25.00
Basic Service with Premium Channels	$20.00 - $25.00

Digital Satellite System

The Digital Satellite System (DSS) is the answer to the variety problem for rural areas of the Midwest. Along with receiving networks such as CNN, C-Span, Headline News, CMT, MTV, and The Weather Channel, the subscriber gets CD-quality audio channels, too. The suburban and rural areas of the Midwest states will enjoy this system. For the same price as premium cable TV, the subscriber will have more channels to watch, and own the equipment. Although DSS provides the customer with 150 channels, the owner still needs a regular antenna to receive local VHF and UHF broadcasts. Typical rates are as follows:

Digital Satellite System Rates

Service	Monthly Rate
Basic Service	$15.00 - $20.00
Basic Service with Premium Channels	$30.00 - $35.00

West Region

The West Region, shown in *Figure 2-4,* includes the Mountain and Pacific States: Alaska, Arizona, California, Colorado, Hawaii, Idaho, Montana, Nevada, New Mexico, Oregon, Utah, Washington and Wyoming

Figure 2-4. West Region of the U.S.
Source: U.S. Bureau of the Census, Statistical Abstract of the United States: 1994 (114th Ed.)
Washington, DC, 1994

Hardline Cable TV System

In the West, the most basic programming service (often referred to as 'tier one') provided by hardline cable TV systems (CATV) includes channels such as CNN, C-Span, Headline News, CMT, MTV, The Weather Channel, and your local VHF and UHF stations. The available number of premium and pay-per-view channels will vary throughout the region. The greater the density of population, the more likely a larger amount of premium and pay-per-view channels will be available. For example, Los Angeles, with a greater concentration of people, will have more specialized channels available than will Clovis, New Mexico. Typical rates are as follows:

Hardline Cable TV Rates

Service	Monthly Rate
Basic Service	$15.00 - $20.00
Basic Service with Premium Channels	$21.00 - $31.00

Home Satellite System

Due to their large size, the majority of home satellite system owners will be in less-populated areas (rural or spacious suburbs). Home satellite systems allow the owner to receive over 250 different channels from multiple satellites. CNN, C-Span, Headline News, The Weather Channel, and the like, can all be received with the home satellite system. If the viewer is eager to see a showing of a network program, he can redirect his satellite dish to an 'East-coast' satellite to pick up the show before anyone else in his area sees it on regular TV. Western viewers have the greatest amount of time to see an early showing of their favorite programs. Typical rates are as follows:

Home Satellite System Rates

Service	Monthly Rate
Basic Service	no charge
Basic Service with Premium Channels	$7.00 - $15.00

Wireless Cable TV System

Wireless cable TV (MMDS) is available in limited areas of this region. In those areas where wireless cable TV is available, standard 28-channel service is offered. This basic service includes channels such as CNN, C-Span, Headline News, CMT, MTV, and The Weather Channel. Premium channels are available, but far fewer than with hardline cable TV — perhaps as few as three or four. HBO, Cinemax, and the Disney Channel are the typical premium channels available on wireless cable TV. Multiple HBO and multiple Cinemax channels are rare on this service. Typical rates are as follows:

Wireless Cable TV Rates

Service	Monthly Rate
Basic Service	$9.00 - $15.00
Basic Service with Premium Channels	$25.00 - $30.00

Digital Satellite System

The Digital Satellite System (DSS) is the answer to the variety problem for rural and sparsely-populated areas of the Western states. Along with receiving networks such as CNN, C-Span, Headline News, CMT, MTV, and The Weather Channel, the subscriber gets audio channels as well. The suburban and rural areas of the Western states will enjoy this system. For the same price as premium cable TV, the subscriber enjoys more channels and ownership of the equipment. Although DSS provides the customer with 150 channels, a regular TV antenna is required to receive local VHF and UHF broadcasts. Typical rates are as follows:

Digital Satellite System Rates

Service	Monthly Rate
Basic Service	$15.00 - $20.00
Basic Service with Premium Channels	$30.00 - $35.00

CHECK POINTS — WHAT ARE YOU CONTRACTING FOR?

As programming costs vary from region to region, so do the costs of specialized installation 'extras.' Most of these 'extras' are required unless you live in a log cabin and can simply drill a hole through an outside wall and connect it up directly to your television set! Chances are you have several TV sets and at least one stereo system to which you would like to connect your video system.

Whether you are installing hardline cable TV, wireless cable TV, or one of the satellite systems, you will probably need to have some internal cable runs to wire up multiple sets. Don't let unexpected extra costs and installation time spoil the pleasure of installing your new video entertainment service! Ask plenty of questions from your cable service provider or retailer of your new satellite system before scheduling installation.

Pre-planning where in your home your cable drops (installation locations) will be will result in a much smoother installation, whether you have it done professionally or elect to do it yourself. Also, find out what do-it-yourself projects are within your abilities and regulations. Hardline cable TV companies must ensure that their installations meet minimum technical standards, and they may require that you use specialized cable, connectors, and accessories to maintain these standards. Here are the definition of several types of installations that the installer may ask you about.

Wall Drop

A wall drop is a type of installation where a coaxial cable is routed down the inside of a wall. This kind of procedure is necessary if you wish to add a new cable TV outlet to a room. Because most cable TV companies have an installation charge for wall drops, always ask how much the wall drop will cost before contracting the cable TV installer for the job. Call the cable TV company and inquire ahead of time. If the cost meets with your approval, then schedule an installer to do the job for you. It is recommended to let a cable TV installer do the work for you when cost is not an issue. If you believe the cost is too high, you can consider installing the wall drop yourself and save money (see Chapter 7).

Note: Depending on which state you live in, there may be government regulations that stop some cable TV companies from billing a subscriber monthly for additional outlets. The companies that are regulated cannot charge monthly rent on two or more outlets. However, many of these companies make up for lost profit by charging rent for converter boxes and remote controls. Remember, if you wish to install your own outlets, turn to Chapter 7 for step-by-step instructions. You can save money by doing it yourself.

Pre-Wiring and Additional Outlets

Pre-wiring a home for cable TV service is another area you can contract for service. Many new home builders have discovered the benefits of pre-wiring. It costs very little to pre-wire a home in the framing stage and makes for a big selling benefit to the homeowner. By installing cable TV outlets in the home ahead of time, the owner forgoes the extra cost and time delays of waiting for the cable TV company to install them.

A similar money-saving idea is to add additional outlets yourself. If you already have cable TV and you want a new outlet in a room, consider installing one yourself! It is legal to add extra outlets to your home. However, it is not legal to connect a cable from your home to the cable company's tap without paying for service.

King for a Day

Today, the consumer is king. You have the right and the choice to do any of the previously mentioned projects yourself. Or, you can arrange to have a professional do the work and provide materials for you. It's your choice! The cable company does not have to perform all the functions associated with doing a wall drop, installing additional outlets, or pre-wiring a home for cable TV. If you have the time, tools and expertise, you can choose to do it yourself.

DO-IT-YOURSELF PROJECTS CAN SAVE MONEY AND ARE AS EASY AS 1-2-3

Don't Give Away Muffy!

Did you know that the cable TV companies now charge you a service fee if you accidentally damage your coaxial cable and they have to make a house call? Suppose your sweet little dog, Muffy, chewed the coaxial cable in two and your cable service is interrupted. Your local cable TV service department will charge you a rate of about $20 per hour for repairs! This varies slightly from region to region. Previously, most repairs were made at no charge to subscribers.

Why not do it yourself? Recent cable regulations allow you to save money by following easy step-by-step installation and repair instructions. Why pay the cable TV company when you can do the repairs yourself? You can save time and money and think of the satisfaction of knowing that *you* did the work! In succeeding chapters, we'll cover video installation and repair. With this information, you should be able to repair your cable if you accidentally damage the coaxial cable in your home.

WHAT'S NEXT?

In the next chapter, we'll look at each of the TV video system components and discuss their specifications.

Chapter 3
Typical System Components

A Bit of the Basics

In this chapter, we discuss our four TV video program sources in a more technical way. This should help you better understand the systems and give you a background for the information in later chapters. We introduce some technical terms, but we'll explain each one as we go. (There are additional definitions in Chapter 7 related particularly to installations. You should refer to these to gain an even better understanding of video system components.) We are not trying to make you a technical expert, but we are trying to give you enough basic facts and system details to allow you to talk confidently about your system. This will help you better communicate with the sales people or technicians you talk to in the process of establishing your TV video service.

COAXIAL CABLE: THE COMMON THREAD

One of the most important components common to all our video system, and probably the one most often overlooked, is the coaxial cable. (Some *techie* types just call it *coax*.) In most cases, the coaxial cable carries only the RF signal from the cable company, or one of the other video sources, into your home and eventually to your TV or VCR. (RF stands for "radio frequency," which is a commonly-used technical term that describes all electro-magnetic communications signals including radio, television, and data transmissions.)

In some systems, however, the coaxial cable also carries low-voltage electrical current used to power devices along the cable. For example, in a large-dish satellite system, the coaxial cable carries power and control signals from the receiver in your house to the amplifier on the dish, in addition to carrying the video and audio signals from the dish to the receiver. Coaxial cable can also carry control signals that allow the cable company to communicate with your cable decoder box in your home. Coaxial cable is truly the heart of all video systems.

The most common type of coaxial cable used to bring signals to your home and distribute the signals within your home is RG-6. (RG-59 cable was used previously, but has been replaced by RG-6 because RG-6 has less signal loss, especially at higher frequencies.) Both RG-6 and RG-59 cable have an *impedance* of 75 ohms. Impedance is a complex technical subject. It isn't necessary to have an understanding of impedance beyond knowing that impedances must be equal or "matched" throughout a system. A mismatch in impedance results in inefficient transmission of the signal and a poor-quality picture. Follow our installation instructions faithfully, and you won't have to worry about mismatched impedances!

Occasionally, a cable TV company uses a special, larger-diameter cable between the tap on its feeder line and your house. It is called RG-11, and it is used when a home needs an especially strong signal to service more than four TVs or other receivers. RG-11 can also be used when the cable run from the feeder line to the home is unusually long. In any case, this is something that only the cable company can do, but if you're having a problem with signal quality because you have several TVs, you know what to suggest to the cable company. They'll think you're a pro.

Because coaxial cable performs such an important job, it is smart to use the best quality cable available. Besides, there's only a few cents-per-foot difference in the cost of ordinary cable and the very best. For longer cable runs (in your walls and attic) we recommend RG-6QS. This cable has a quadruple shield to protect your signal for almost any kind of interference. For short cables between the components within your system (TV, VCR, and so forth) you can use standard RG-6 cable. You can buy bulk quantities of RG-6 in black or white, or you can buy various lengths of pre-made cables. All of these cables are low-loss and are UL-recognized for any of the installation examples presented in this book. Contact your local electronics parts distributor for information about any special requirements for cable in your area.

HARDLINE CABLE TV SYSTEM COMPONENTS

There are numerous components that make up a typical hardline cable TV system. After all, for the signal to get to your home from its original source at a network, it has to travel almost 50,000 miles. That takes some doing. You never see most of the components, and you don't have to understand the technical stuff. We'll just give you a quick summary so you can appreciate the complexity of the system.

Uplinks and Downlinks

The signals for your cable TV programs begin at the network studio—either directly from a camera or from a taped source. From there, they are sent to an earth-station "uplink." The earth-station *uplink* uses a large satellite dish to transmit the signals to one of the many communication satellites. These satellites are about 22,300 miles above the earth's equator in what is called a *geostationary* orbit. This means that the satellites are moving at exactly the same speed as the earth's rotation. From the earth, the satellites appear to be sitting still.

From the satellite, the signal is relayed back toward the earth (on a different frequency so it won't interfere with the uplink signal). These signals are the "downlink" and are picked up on satellite dishes at the cable company's earth-station *downlink*. Signals from many of the satellites can be received anywhere in the United States, but satellite signals do have *footprints* as shown in *Figure 3-1*. In other words, the signals are focused on a certain area, and the signals get weaker the further away from that area you get. That's why in some parts of the country you need a larger satellite dish to *collect* enough signal to produce an acceptable picture. Cable companies use very large dishes (up to 30 feet in diameter) to assure the best quality picture and sound. After the signals are received by the downlink, they are sent to the next component in the hardwire cable system: the "head end."

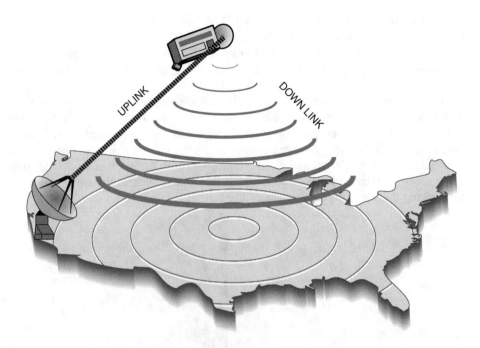

Figure 3-1. Satellite Footprint—The Satellite Signal is Strongest at the Center of the Footprint and Decreases as the Distance from the Center Increases.

The Head End

The *head end* is the general name for a room full of components at your local cable company. The head end receives the signals from the downlink, fine tunes them and sharpens them, converts them to the channels you receive on your TV, and sends them out through the various trunks and feeder lines of the cable company's distribution system on their way to your home or business.

At the head end, there is a separate receiver for each channel that is received and sent out to you. If a cable provider in a town offers 60 channels for viewing, they have 60 separate receivers! This includes receivers for each local VHF and UHF station.

Why are all these receivers necessary? They are designed to produce a perfect image and perfect audio for distribution throughout the system. Remember, the receiver in your TV only has to provide pictures and sound for *you*. The cable company's receivers provide pictures and sound for *thousands* of customers.

By the way, have you ever noticed while surfing through the cable channels, that the sound of some stations is louder than that of others? The head end is probably the source of the problem. The audio output levels of the different receivers are not properly matched.

Distribution Equipment Along the Way to You

The head end distributes the signals to a trunk, which distributes the cable signals through a network of trunk lines and feeder lines. *Figure 3-2* shows a typical way to distribute cable signals. The feeder lines are strung between high-voltage power line poles.

One of the most serious obstacles the cable company faces in getting their signals to their customers is *signal loss*—the further the signal goes from the head end, the weaker it becomes. The coaxial cable has resistance that weakens the signal and so do the taps that are inserted along the cable to allow connections to your home or office. It's just like water running through a garden hose. A long hose offers more resistance to the water, and if there are holes in the hose to let out water along the way, there is less water the further it goes.

HIGH-VOLTAGE
POWER LINES

FEEDER
LINE

LINE
EXTENDER
(AMPLIFIER)

Figure 3-2. Feeder Line and a Line Extender (Amplifier) Mounted on a Utility Pole. Danger! This Pole Also Carries High-Voltage Power Lines!

The solution for signal loss is amplifiers called *line extenders* or LEs. One of these is inserted every so often in the feeder lines to ensure the proper signal strength. Engineers, technicians, and cable system installers use a term called decibels (dB) to describe signal strength. (Typically, the signal strength in feeder lines is maintained at 40 - 45 dB.) The power for these LEs comes from external sources. As we suggested earlier, the power could be sent along with the TV signals in the coaxial cable, but the cable company doesn't do this in an effort to keep the TV signals as *clean* as possible. *Figure 3-3* shows a close-up view of a feeder line, line extender, and a "line tap." The *line tap* is the next component in the hardline cable system.

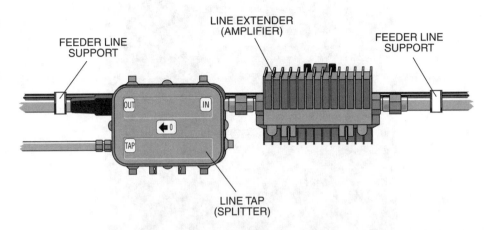

FEEDER LINE
SUPPORT

LINE EXTENDER
(AMPLIFIER)

FEEDER LINE
SUPPORT

OUT

IN

TAP

LINE TAP
(SPLITTER)

Figure 3-3. Close-Up View of a Feeder Line with a Line Extender (LE) and a Line Tap.

Tapping the Signal

The *line tap* (sometimes just called a *tap*) is the device installed on a feeder line that splits the signal and provides an output for the cable company to connect the signal to your home or office. Two different types of taps are shown in *Figure 3-4*. A tap is found in one of two places. If the feeder lines are above ground, the tap is located on a utility pole in or near your yard. If the cable system uses underground feeder lines, the tap is inside a "pedestal:" a metal box on the exterior of the building. *Figures 3-5* shows a pedestal for an apartment complex. A tap can have several outputs, and each output can provide service for a typical home or office. The signal strength at each output should be from 6 to 10 dB.

Figure 3-4. Line Taps for Feeder Lines. A 4-Way Tap is Shown on the Left and an 8-Way Tap is Shown on the Right.

Figure 3-5. Connections to an Underground Cable Line are Made in an Accessible Housing Called a Pedestal.

The Cable Drop to Your Home

The "cable drop" is the name installers use for the coaxial cable that goes from the tap on the feeder line to your home. As we mentioned earlier, this cable is usually high-quality RG-6 cable. And, if you have more the four TVs or your house is a long way from the feeder line and tap, the cable company might install a more efficient RG-11 cable to minimize signal loss.

Grounding the System

One of the most important components of any video system that receives signals from outside the home is a *ground block*, shown in *Figure 3-6*. It helps protect you and your inside components from high-voltage static electricity that can build up gradually or occur as a spike from a nearby lightning strike. The ground block consists of three connections: one for the input from the *cable drop*, one for the output to the coaxial cable going into your house, and a third for the wire that leads to a grounding rod or to the metal conduit for your home's electrical meter. It is very important that your grounding system comply with local regulations and codes. Typically, a 8-foot grounding rod connected with 8-gauge wire is a minimum requirement, but check to be sure.

In addition to being an important safety feature, the ground block serves another purpose. It is the dividing line between the part of your cable TV system that belongs to you and the part that belongs to the cable company. This is an important division when it comes to maintaining or adding accessories to your system, and we'll talk about it more in other chapters. For now lets get back to the components.

Figure 3-6. Ground Block. This is an Essential Component in Any Video Installation. It Protects Against Potentially-Damaging Power Surges.

THE COMPONENTS IN THE HOUSE

There are four main components on your side of the ground block, shown in *Figure 3-6*, in a typical hardline cable TV system. The components are the house drop, the wall-plate outlet(s), the cable converter box(es), and the TV(s).

The House Drop

The *house drop* is nothing more than the coaxial cable that begins at the ground block and is routed through walls, crawl spaces, and/or the attic to a video system wall-plate outlet near your video system. The house drop should be high-quality RG-6 cable.

The Video System Wall-Plate Outlet

The video system wall-plate outlet is similar in size and shape to other outlets in your home (electrical outlets, telephone outlets, etc.), but it has a different type of connectors. The screw-on connectors used on the ground block, coaxial cable, the wall-plate outlet, and almost all video equipment are called "F connectors." Typically, each end of a coaxial cable has a male F connector, and the inputs for all other components and accessories have female F connectors.

Each wall-plate outlet has two F connectors—one inside the wall and one outside. The house drop connects to the one inside the wall, and the coaxial cable leading to your video system connects to the one outside. There are two advantage of using wall-plate outlets instead of running the house drop directly to your video system: it looks neater, and it allows you to easily lengthen or shorten the cable if you move or rearrange your video system.

The Converter Box

The next component in most hardline cable systems is the converter box, also called the cable decoder box. An older style converter box is show in *Figure 3-7a,* and a newer style converter box is shown in *Figure 3-7b*. The rear view in *Figure 3-7c* shows the female F connectors.

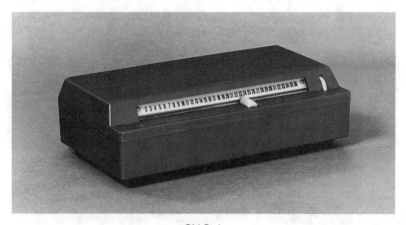

a. Old Style

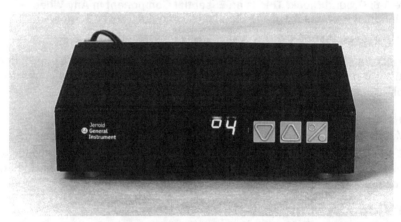

b. Newer Style – Front View

c. Newer Style – Rear View

Figure 3-7. Typical Top-of-TV Converter Boxes Furnished by Your Local Cable Company. They Convert Frequencies and Process Scrambled Programs.

There are two main reasons the cable converter is required. First, most cable channels are on frequencies that a standard TV can't receive. Second, most cable companies scramble some or all of their channels. Here's what each of these reasons means to you.

Cable Channel Frequencies

The signals in a cable system are grouped into several frequency bands. Cable channels 2 – 13 are exactly the same as off-the-air VHF broadcast channels 2 – 13. So if your cable company doesn't scramble these channels, you can receive them directly on any TV without a cable decoder box.

However, all other cable channels use different frequencies than broadcast TV. This can be confusing because the upper cable channels and UHF broadcast channels use the same channel numbers (14 and above), but *they are not the same frequencies*. Cable channels 14 and above are divided into several bands, such as mid-band, hyper-band, and super-band, but these are seldom mentioned except in very technical conversations.

If you have a cable-ready TV and channels 14 and above are not scrambled, you can receive them without a cable decoder box. But the reality is, fewer and fewer cable channels are *clear to view* (not scrambled).

Note: It's important to note that many of the cable frequencies are shared by wireless communications such as business, public-service, and amateur two-way radio. All the more reason to use properly shielded coaxial cable to keep these signals and your cable signals from interfering with each other.

Scrambled Channels

If most or all of your cable channels are scrambled, it doesn't matter much if your TV and VCR are cable-ready or not. The cable signals must be decoded by the cable box before they get to your TV. When you select a channel at the cable box it descrambles it, converts it to a standard TV channel (usually 3 or 4), and sends it to your TV. The limitation here is that the cable box can descramble and output only one channel at a time. So regardless of how many channels your cable company offers, you are limited to recording and viewing one at a time. There are several ways to improve this situation, and we'll discuss these in more detail in Chapter 8.

Typical Satellite System Components

There are several components that make up a typical large-dish satellite system. As in hardline cable TV, home satellite system signals start at various network studios. These signals are sent to earth-station uplinks where they are beamed to satellites in geostationary orbits. From the satellites, the signals are relayed to an earth-station downlink. The difference in a home satellite system is that the earth-station downlink is the satellite dish in your backyard, not the dish as your local cable company. At your satellite dish, the signals are collected, amplified, and converted by some rather sophisticated electronics and sent into your home through coaxial cable. Inside, a satellite receiver converts the signals a final time to a channel your TV can receive.

There are four main components in a home satellite system: the satellite dish (also called a downlink satellite dish), the LNBF, the house drop, and the satellite receiver. Each one plays an important role in bringing pictures to your TV. Let's talk about the satellite dish first.

The Satellite Dish

Your home satellite dish is a large parabolic reflector. In case you have forgotten your basic geometry, here's a little refresher! The shape of a parabola is the bottom quarter of a sphere. The satellite dish is shaped this way to reflect, or *focus,* a signal on a single point. The larger the dish, the greater the amount of signal it can reflect and focus. Satellite dishes are available in several sizes from 5- to 10-foot diameters. A typical satellite dish is shown in *Figure 3-8.*

Figure 3-8. A Home Satellite System Large Parabolic Dish Antenna.

Generally, a larger dish provides stronger signals and a better picture. Your distance from the equator and your position within the satellite's footprint are primary factors in determining the size dish your need. In many areas, a 5-foot dish provides adequate reception. In the extreme reaches of the U.S., such a Northern Maine, Southern California, the Pacific Northwest, and Southern Florida, a 10-foot dish might be required.

To get the most from your home satellite system, the dish should have a *dish positioner* attached. This is a motorized device that, when properly aligned, allows you to quickly aim the dish at various different satellites using controls built into your satellite receiver. We'll tell you more about that when we discuss the receiver.

Different Satellite Frequency Bands

Home satellite dishes are used to receive two bands of frequencies—C-band and Ku-band. C-band is most prevalent and provides up to 24 *transponders* from each satellite. (Transponder is the name used for video channels on a satellite.) C-band frequencies are in the range of 3.7 to 4.2 GHz. There are fewer Ku-band satellites, and receiving their signals requires a little more precision in aiming the dish than is required for C-band. But many premium channels are available on Ku-band, and these satellites transmit stronger signals, so your can use a smaller dish if you are receiving only Ku-band. Ku-band signals fall in the 11.7 to 12.2 GHz range.

With either satellite frequency band, the signals captured by the satellite dish are focused on the next component in the system—the LNBF.

The Low-Noise Blockconverter/Feedhorn (LNBF)

Low-noise blockconverter/feedhorn (LNBF) is a long name for an important and complex component in your home satellite system. It is the device mounted on the boom out in front of the dish, and it is actually an integration of three separate components that were used in early home satellite systems—the feedhorn, the low-noise amplifier (LNA), and the block converter. The LNBF collects the signals, amplifies them by up to 100,000 times, and converts them to a lower frequency range (usually 950 to 1450 MHz). These lower frequencies are more easily sent through coaxial cable to the satellite receiver without excess signal loss. You must have separate LNBFs for C-band and Ku-band, or have a special combination C/Ku LNBF, to receive both signal bands using the same dish.

The House Drop

The house drop for a home satellite system is similar to any other house drop. It simply carries the signal inside your house from the satellite dish. As with any other house drop, you should insert a ground block just before the entry point into your house. And, as always, be sure that your grounding rod and connecting wire meet all local regulations.

The Satellite Receiver

The last major component of the system is the satellite receiver. It receives signals from the satellite dish through the house drop and your video system wall-plate outlets. Its primary function is to select one of the satellite channels and convert it to a channel you can view on your TV (usually 3 or 4). Today, most receivers are IRDs (Intergrated Receiver Descramblers). They contain the Video-Cipher II™ descrambler necessary to receive most scrambled channels. You must contact a satellite service provider and have your descrambler activated before you can view scrambled channels.

The satellite receiver communicates with the dish to select which *polarized* signals are received (horizontal or vertical). *Polarizing* is a sophisticated way of transmitting two channels in the frequency range normally used for only one channel. This technique doubles the number of channels available from each satellite. The receiver also lets you tune to any of the hundreds of separate audio-only channels that are included in the satellite signals.

The satellite receiver does have one limitation. Similar to a cable decoder box, it can select only one channel at a time. However, you can purchase a second receiver (a basic, low-cost model) to allow independent channel selection on two TVs. Of course, you can still only aim the dish at one satellite at a time, and both selected channels must have the same polarity. Even so, you'll usually have 12 channels to choose from.

The satellite dish system has another limitation; it can't receive local broadcast channels. If you want to receive your local broadcast stations, you'll need a standard TV antenna. Most satellite receivers have an input for a standard antenna and a built-in A/B switch that lets you easily select between satellite and local programs.

The Dish Positioner

One of the most convenient features of the receiver is the dish positioner controller. Once you program the position of each satellite into the receiver, you can move the dish from one satellite to another as easily as changing channels on your TV. And in most cases you can operate the positioner by remote control, so you don't have to leave the comfort of your chair or couch!

Typical Wireless System Components

In many ways, a wireless cable TV system is the same as a hardline cable system. The programming originates from the same place as it does for other video services. A network studio sends its signals to an earth-station uplink; the signals are beamed to a satellite orbiting in space; and the signals are relayed back toward Earth and are received by the earth-station downlink at the wireless cable TV company. From there, the signals are processed at the head end—all as we described for hardline cable systems earlier in this chapter. But this is the point where the wireless system is different.

From the head end, the signals are transmitted from the top of a high structure in the area, such as a tall building, mountain top, or a transmitting tower, via microwaves to subscribers. At each subscriber's home, the signals are received by a small microwave antenna. From there, a coaxial cable carries the signals to the ground block, and the house drop distributes the signal to the video system wallplate outlets where connections are made to the TV and other video system components. As you can see, from the antenna, it's very similar to all the other TV video systems.

There are five primary components in a typical wireless cable TV system: an MMDS (Multichannel Multipoint Distribution System) microwave antenna, an off-the-air TV antenna (optional), a house drop, a special power injector, and a converter box.

The Microwave Antenna

The microwave antenna for a wireless cable system is usually mounted on a small tripod installed on the subscriber's roof. (This antenna must have a line-of-sight to the transmitting antenna.) If you want to receive local broadcasts, you'll need an off-the-air TV antenna, too; you can mount it on the same mast as the microwave antenna.

The microwave antenna, shown in *Figure 3-9,* actually consists of two components—the antenna, and the low-noise amplifier/down converter (LNA/DC). They might look pretty simple to you, but they are very sophisticated pieces of equipment. The antenna itself has a high gain (about 32 dB), which means that it captures a lot of the transmitted signal. It is a reflector-type antenna, and works very much like a parabolic satellite dish. After the signals are captured by the antenna, they go to the 33-channel, low-noise amplifier/down converter (LNA/DC), shown in *Figure 3-10,* where they are amplified and stepped down from their original 2.5 to 2.686 GHz range to a 222 - 408 MHz range. The conversion to a lower frequency range is important for two reasons—it allows you to use standard coaxial cable for the house drop, and it allows you to combine the VHF/UHF broadcast signals and the wireless cable signals so that you have only one coaxial cable leading into your house. To understand how this works, you need to know the frequency ranges of all the signals.

Figure 3-9. High-Gain Multipoint Microwave Distribution System (MMDS) Wireless Cable Antenna. The Details of the Feedhorn are Clearly Shown in the Side View.

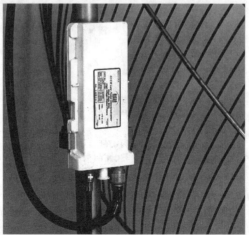

Figure 3-10. The Low-Noise Amplifier/Down Converter Box Behind the MMDS Antenna Converts the Signals so Regular Off-the-Air Antenna Signals Can be Combined with Those of the Wireless Cable System and Sent Down One House Drop.

VHF channels 2-13 are in the range of 54 to 216 MHz. UHF channels 14-83 are in the range of 470 to 890 MHz. The microwave signals are received in a range of 2.5 - 2.686 GHz, but are converted by the diplexer to a range of 222 - 408 MHz. So when all the signals are sent through the same coaxial cable, the wireless cable signals fit in the gap between the VHF and UHF signals, and there is no interference.

A Special House Drop

As you probably know by now, the house drop is a 75-ohm cable that brings a signal into the home. The house drop on the wireless cable TV system is special because it serves two purposes. As in our other TV video system, it distributes the signals within your house. But it also carries power to operate the components on the antenna. Here's how!

The house drop runs to a power injector inside the home near the television set. The power injector, shown in *Figure 3-11,* is connected to an ac wall outlet through an ac adapter, and it *injects* 24-volts DC up the cable to the antenna. The electricity powers the LNA/DC and eliminates the need for a separate power cable. Special circuits at the power injector and at the microwave antenna separate the DC current and video signals.

Figure 3-11. Power Supply for MMDS LAN/DC. The Small Silver Box at the Left is the MMDS Power Injector and the Black Box at the Right Is the AC-to-DC Adapter.

The Converter Box

The signals from the LNA/DC go to a converter/decoder box that is on or near your video system components. A typical converter box is shown in *Figure 3-12a.* The converter box descrambles the signals for normal viewing and converts them to a standard TV channel (usually 3 or 4). Notice also that the converter box has separate audio and video output jacks as shown in *Figure 3-12b*. These jacks allow direct connection of the audio and video signals to your VCR or TV set.

a. Front View

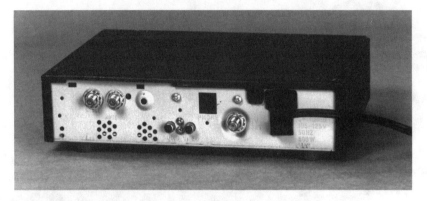

b. Rear View

Figure 3-12. Front and Rear Views of a Typical MMDS Converter Box.

DIGITAL SATELLITE SYSTEM COMPONENTS

The latest technological breakthrough in TV video systems is the Digital Satellite System (DSS). Although its antenna is much smaller than the more familiar large-dish home satellite antenna, this system possesses many similar components. The network studios send signals to a geostationary satellite from an earth-station uplink. Then, the satellite relays the signals to a miniature earth-station downlink—the 18-inch satellite 'downlink' dish on each DSS subscriber's home.

Digital Satellite System downlink signals are much stronger than the C-band downlink signals used by large-dish home satellite systems. The combination of these more powerful signals and the higher Ku-band frequency range (10.95 to 12.7 GHz) allows the dish to be much, much smaller than those used in older, large-dish systems. In fact, the DSS downlink signals are seven times stronger than the signals from the older satellites! And the fact that the signals are digital, instead of the traditional analog, means that pictures and sound are received without the video *noise* and static that sometimes affect large-dish reception. This makes DSS audio and video quality comparable to that of a laser disc.

There are five main components in a Digital Satellite System—an 18-inch satellite dish, a 75-ohm coax cable house drop, a DSS receiver, a telephone jack, and an off-the-air TV antenna (optional). The following is the typical route signals travel through the components of the DSS.

The DSS Dish

The digital signals are captured by the 18-inch satellite dish, which is mounted on the subscriber's home. In contrast to the moveable dish of the large-dish system, the DSS dish is fixed in one position. It receives signals from only one satellite instead of several. The signals are focused on the antenna's feedhorn, which also contains a low-noise amplifier and down converter. A typical DSS dish and receiver are shown in *Figure 3-13.*

Figure 3-13. Digital Satellite System (DSS) Satellite Dish, Receiver, and Remote Control.

The House Drop and Receiver

A house drop brings the signals into the house via the ground block and connects to your video system wall-plate outlets. From there, a short cable completes the connection to the DSS receiver's input. The DSS receiver changes the digital signal into audio and video signals for your TV.

A Telephone Line for Pay-Per-View

There is a telephone line connection on the back of the DSS receiver that is necessary for automatic requests and billing for pay-per-view programs. When you watch a pay-per-view program, the request is recorded in the DSS receiver's memory. The receiver automatically calls a centrally-located computer and forwards the necessary information to generate a bill.

Standard TV Broadcasts

As with the large-dish satellite system and the wireless cable system, if you want to receive local off-the-air broadcasts, you'll need a standard TV antenna with a separate house drop. The cable from this standard antenna is connected to the DSS receiver, and you can use the receiver's built-in A/B switch to easily select between the two signal sources for your TV.

PROGRAMMING FOR THE TV VIDEO SERVICES

Though our four TV video systems are different in many ways, the programming available from each system is similar. It helps to know a little about the various cable-type channels and what is included in *basic* and *premium* service. Here is a brief synopsis.

Basic Service

Basic service typically includes Nickelodeon, ESPN, USA, The Weather Channel, TNT, MTV, VH-1, several superchannels (such as WGN and WTBS), your local broadcast stations and network affiliates (ABC, NBC, CBS, FOX). Premium Channels are not included.

Pay Channels and Premium Channels

Premium channels are ones with movies that have recently completed their first run in the theaters, made-for-cable movie productions, and musical and comedy specials. HBO and Cinemax are good examples of premium channels. They are owned by the same company. If a movie is shown on HBO, eventually it will be shown on Cinemax (or vice versa). Showtime and The Movie Channel have a similar symbiotic relationship. What is shown on one will be shown on the other.

Prime Ticket™ (formerly Home Sports Entertainment) is an all sports channel. Some cable companies offer Prime Ticket with basic service; others do not. Ask your service provider about this one.

The Disney Channel shows Disney-produced programs for children and adults, as well as other movies that have already been exhibited in theaters. Disney full-length movies and cartoons, special events, and concerts are also part of programming on the Disney Channel. High-quality entertainment has been Disney's hallmark. You'll find this quality on their video channel, too.

"Pay-per-view" channels offer the same first-run movies as your neighborhood video rental store, but you don't have to leave home to get them. Specials, such as music concerts and sporting events, are also pay-per-view fare. Usually, you must call to order pay-per-view events. However, if you have interactive (two-way) cable, you may select pay-per-view events using your remote control. Talk about convenience! Instead of paying immediately, you are billed for the movies and special events you order at the end of the month, along with your regular video billing.

Hardline Cable Programming

Hardline cable TV systems offer many different channels. Depending on the system and the city you live in, the cable company can offer a range of from 30 to more than 80 channels. Most often, these systems offer basic, premium, and pay-per-view channels. Your local TV listings (in your local newspaper or TV Guide) will include the cable channel numbers for your convenience.

Home Satellite System Programming

Large-dish satellite systems offer the broadest range of programming to their owners. Subscribers can choose from more than 250 channels. There are many different satellites and many programming providers. Satellite system owners have access to basic programming, premium channels, and pay-per-view channels. Providers include Turner Home Satellite, Netlink, Disney Home Satellite Services, Multichannel HBO/Cinemax, and Playboy, just to name a few. To identify what channels and programming packages are available, contact your satellite system dealer or check the listing of a large-dish satellite guide, such as *Satellite TV Week* or *Orbit.*

Wireless Cable Programming

Wireless cable TV (MMDS) offers the least number of channels of the TV video systems. Wireless cable TV systems offer their subscribers 28 to 33 channels. They are limited because of the narrow frequency range allocated to them by the Federal Communications Commission (2.5 - 2.686 GHz). Like other TV video systems, wireless cable TV offers the subscriber basic service. Also, like traditional cable TV, premium channels and pay-per-view channels can be added to the basic programming for additional charges. To identify what services are available, check with the wireless cable TV operator in your area.

DSS Programming

The Digital Satellite System has the second largest channel selection to offer its subscribers. It offers more than 150 video channels and more than 25 music-only audio channels. The DSS subscriber has access to basic programming, premium channels, and many pay-per-view channels. Currently, there are two primary providers of DSS programming: DirecTV Services and United States Satellite Broadcasting (USSB). Both offer a variety of packages for the cost-conscious to the extravagant. If you want all five HBO channels, you can have them! If you want non-stop sports, you can have this, too! To identify the channels and packages available, contact your DSS dealer or check the listing of a DSS TV guide such as *Satellite Choice.*

SUMMARY

As you can see, you are not limited to traditional cable TV for more varied television programming. Real alternatives to cable exist and more are sure to be developed in the future. Compared to the cost of an evening's entertainment (dinner and a movie, a sporting event, or a concert), premium programming services are a true bargain today. The cost of systems has steadily deceased, as well. If you've ever thought about getting cable or one of its alternatives, there's no reason to wait. As we continue our explanation of video systems, we'll show you how to get the most from whatever system you choose.

Chapter 4
Apartment Installations

INTRODUCTION

Millions of Americans live in multiple dwelling units such as apartments, flats, lofts and condominiums nationwide. People who live in such housing are limited in the choice of the type of TV video system they may have. Apartment dwellers typically have hardline cable TV (CATV) or a Central Antenna Distribution System (CADS). The hardline cable TV is the same cable TV most residential consumers have. CADS uses a central antenna system and local broadcast signals are distributed through a coaxial cable network to individual apartments. CADS is also known as MATV (Master Antenna Television).

In this chapter, we will show the typical components used in an apartment installation, and the common routing methods used when cable is installed in an apartment complex, both hardline and the Central Antenna Distribution System. In addition, we'll show you how to move a cable outlet in an apartment. Even though Chapter 7 covers typical "do-it-yourself" installations in detail, many people living in an apartment might need to change the location of their cable outlets. Maybe the use of a room has changed, or a second outlet is desired in a bedroom as well as the living room. Or there may be a need just to move the TV within a room. All of these cases will be fully illustrated.

HARDLINE SYSTEMS

A typical hardline cable TV system installed in an apartment complex is shown in *Figure 4-1*. The cable signal is delivered to the apartment in a similar fashion as to a single-family residence. The signal leaves the cable company's head-in and is routed to a trunk line. From the trunk line, it is connected to feeder lines that distribute the signal throughout the network. In an area of a network where there is an apartment, an amplifier is inserted in the feeder line to amplify the signal before it is sent to the apartment complex. The amplifier boosts the signal so that it strong enough for all the apartments in the complex. A feeder line from the amplifier is routed to the first apartment building in the complex. The feeder line terminates inside a metal box, called a pedestal, that is mounted on the exterior wall of the apartment building.

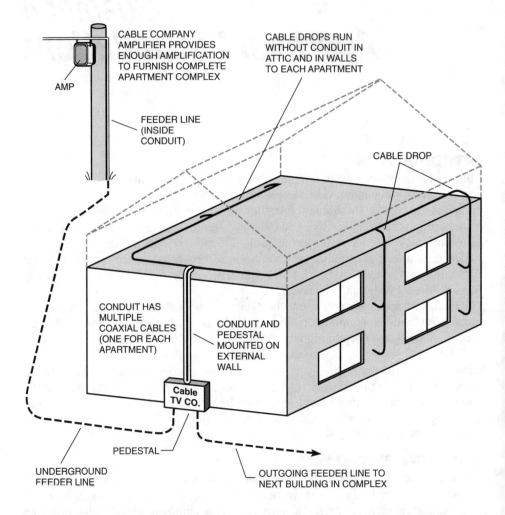

Figure 4-1. The Cable Tap is Located Inside the Pedestal and the Coaxial Cables Feed Through a Conduit to the Attic and Then Drop to Each Apartment Unit (For Apartment Complexes That Were Wired for Cable After Construction).

Apartment Distribution

Inside the pedestal are taps for the various apartments. From the tap, a cable drop runs to each apartment as shown in *Figure 4-1*. There is one outlet and one cable to each apartment. The cable drops from the pedestal are run inside a conduit attached to the exterior wall which then goes through the wall of the apartment building into the attic. In the attic, cable drops run across the attic and inside the walls to the individual apartments. Also, as shown in *Figure 4-1*, an outgoing feeder line goes from the pedestal to carry the signal to the next building in the complex (if there is more than one building).

Conduits

As already stated, in most apartments the cable drops leave the pedestal and are routed to the attic of an apartment through a conduit. A conduit is a protective covering for the cable lines. It is made of metal or plastic and prevents the deterioration and weathering of exposed lines. Conduit has another redeeming quality: it looks nice! A single conduit leading to the attic looks much better than several exposed cables. Depending on its size, many cables may be routed through a conduit.

Cable Outlets in the Apartments

The cable drops inside the walls terminate in cable outlets in each apartment. The outlets have wall-plate cable connectors covering the entrance inside the apartment. The wall-plate connector is shown in *Figure 4-2*. The end of the cable drop has a male F connector that screws onto the cable connector on the wall-plate connector as shown in *Figure 4-2b*. Inside the apartment the cable outlet looks like *Figure 4-2a*, the front view of the wall-plate cable connector. A connecting cable with two male F connectors is screwed to the wall-plate cable connector (*Figure 4-2b)* to carry the cable TV signal to a converter box on top of a TV. Two types of F connectors are shown in *Figure 4-3*. *Figure 4-3a* is a screw-on connector; *Figure 4-3b* is a push-on connector. Cables that are likely to be pulled off, or in high-traffic areas, should have screw-on connectors. The converter box on top of the TV is needed to process scrambled cable channels.

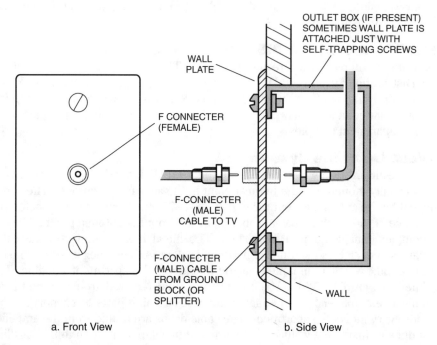

Figure 4-2. A Wall-Plate Cable Connection Commonly Found in Apartment and Home Cable Installations.

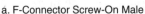

a. F-Connector Screw-On Male   b. F-Connector Push-On Male

Figure 4-3. The Coaxial Cable F Connectors Used for Cable Installations are Usually Screw-On Types. The Push-On Connectors are Usually for Video Equipment.

It should be noted that the typical installation shown in *Figure 4-1* is for an apartment building in which the cable drops and pedestals were installed while the building was being constructed. Many installations are made after the building was already constructed. In that case, all the cable drops would likely be in conduits run on the outside of the building, and cable drops to each apartment would be through exterior walls into each apartment. The cable drop termination is to the same wall plate-cable connector, shown in *Figure 4-2b*.

CENTRAL ANTENNA DISTRIBUTION SYSTEM

In apartment complexes where hardline cable TV is not available, a Central Antenna Distribution System (CADS) is installed. A typical system is shown in *Figure 4-4*. A Central Antenna Distribution System is exactly what the name implies. It is a centralized antenna system which distributes an off-the-air signal to apartments within an apartment complex.

A Close Up of the System

The off-the-air antenna which gathers all the local UHF and VHF stations in your area is usually mounted at the highest point in the apartment complex. The signal collected by the antenna is sent through a 300-to-75-ohm transformer to match the impedance of the 75-ohm coaxial cable that runs from the antenna to a building in the complex that contains a metal box. The metal box is similar to the cable company's pedestal, as used in the hardline system. An amplifier is located inside the metal box which is mounted on the outside of the apartment building. The amplifier, powered by 120 VAC, boosts the signal received from the off-the-air antenna. The amplifier has an equal number of outgoing ports to support a cable drop for every individual apartment. The cable drops are routed from the amplifier in conduits or inside the walls of the apartment in a similar fashion to the hardline system. The cable drops end up connected to a wall-plate cable outlet in each apartment. Of course, there are some apartment complexes that will route the

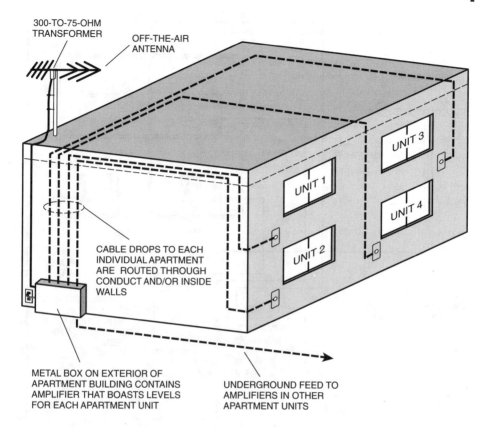

300-TO-75-OHM
TRANSFORMER

OFF-THE-AIR
ANTENNA

UNIT 3

UNIT 1

UNIT 4

UNIT 2

CABLE DROPS TO EACH
INDIVIDUAL APARTMENT
ARE ROUTED THROUGH
CONDUCT AND/OR INSIDE
WALLS

METAL BOX ON EXTERIOR OF
APARTMENT BUILDING CONTAINS
AMPLIFIER THAT BOASTS LEVELS
FOR EACH APARTMENT UNIT

UNDERGROUND FEED TO
AMPLIFIERS IN OTHER
APARTMENT UNITS

Figure 4-4. Off-the-Air Central Antenna Distribution System Physical Layout (The Weak RF Signal from the Antenna is Amplified so That It Provides Strong Signals to All Apartment Units).

cable drops in conduits on the outside of the building. This method of routing cable is called *wrapping*. It literally means to wrap the cable drops around the exterior of the apartment building.

Schematic Diagram

To better understand the interconnection of the Central Antenna Distribution System, *Figure 4-5* is a schematic drawing of the system showing the distribution amplifier cable connections to the wall plate outlets in the apartments. The distribution amplifier is an eight-way amplifier, but it is connected to only four units. Note the output that feeds to the amplifier in the next building of the complex, and provides for distribution throughout the complex.

Obviously, if the amplifier develops trouble, the whole building will lose TV reception. Off-the-air reception is also dependent on atmospheric conditions, the line-of-sight path to and the distance from the station antennas. Sophisticated systems will have several antennas pointed in various directions, each with its own preamplifier for optimum reception.

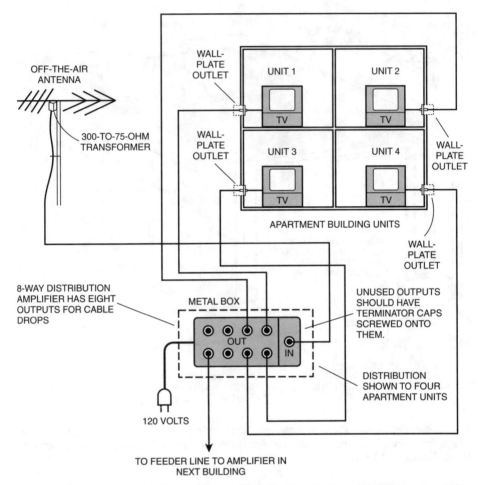

Figure 4-5. Schematic of Central Antenna Distribution System. Eight-Way Amplifier Shown Distributing Signal to Four-Apartment Unit.

QUICK AND EASY OUTLET ADDITIONS

In some homes and apartments, the living room is the only room that has an outlet; however, many people prefer to have an outlet in the bedroom. Or the location of an outlet within a room is not the ideal location for a TV. It would be convenient if a cable extension could be added so that the TV can be relocated. Both situations will be examined in closer detail.

In either case, there are two ways to relocate the outlet. You can have the cable TV company do it for you or you can do it yourself. If the cable company does the job, you will pay an installation fee, plus parts, and have the inconvenience of scheduling a time when the cable company can do it. The alternative, doing it yourself, is cheaper and easier than you think. Before proceeding with a do-it-yourself installation, make sure you are not violating your lease agreement by making any permanent changes to your apartment's cable wiring.

Adding a Cable Extension

Figure 4-6 shows a cable extension that has been added to move the location of a TV outlet to the other side of a doorway. Here are the easy steps necessary to do it:

1. Make an approximate measurement of the length of cable required from the present cable outlet to the new TV location. Some cable assembly lengths (cable with connectors installed) are available at your local electronics parts store. Some store personnel can also cut a custom length and have the cable made up for you, if you desire. Choose the cable for your application. The cable should have male F connectors on each end, either the screw-on or push-on type shown in *Figure 4-3*. The screw-on type is preferred to make for the most rugged electrical connection. The maximum length you should use is 100 feet. Anything over this length will cause undesirable signal loss.

Typical Coaxial Cable Extensions Available		
Cable	Length	Description
A	16'	75-Ohm Coax with Male F Screw-Type Connector
B	25'	Same as A
C	50'	Shielded 75-Ohm Coax with Male F Screw-Type Connector
D	100'	Same as D

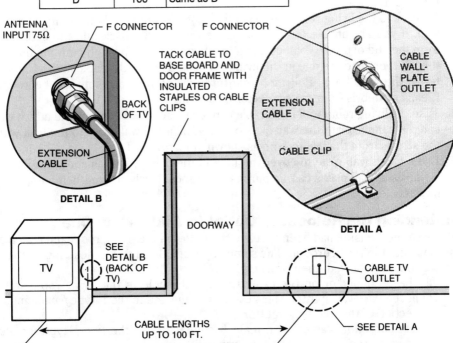

Figure 4-6. Adding a Coaxial Cable Extension to Extend the Location of an Outlet (Cable Lengths Should Not Exceed 100 feet)

2. Connect one end of the cable to the wall-plate cable outlet, and dress the cable along the baseboard as shown in Detail A of *Figure 4-6*.
3. Tack the cable to the baseboard and door frame, dressing it neatly along the way. Tack it with insulated staples, staples from a staple gun using T-37 or T-25 staples, or cable clips.
4. Connect the other end of the cable to the 75-ohm antenna connection on the rear of the TV as shown in Detail B of *Figure 4-6*. That's all there is to it!

Relocating an Outlet

The apartment used as an example has a bedroom behind the living room as shown in *Figure 4-7a*. The cable outlet in the living room is to be relocated to the bedroom. Here are the easy steps to accomplish this:

1. Remove the wall plate by removing its two screws.
2. Pull the wall plate and cable connected to it out from the wall as shown in *Figure 4-7b*, and remove the cable from the wall-plate outlet by unscrewing the F connector.
3. Take a screwdriver (6 inches or longer), and stick it through the hole in the wall where the wall plate was, and from which the cable still extends. With the screwdriver and a hammer, punch out a hole in the wall of the adjacent bedroom, as shown in *Figure 4-7c*.
4. After you have made a hole in the wall, go into the bedroom and widen the hole with your screwdriver. Make it large enough so that a cable may be pulled through it.
5. Go back to the living room and push the cable through the hole in the bedroom wall. If you are unable to push the cable through, go back into the bedroom, form the end of a wire coat hanger into a hook, push it through both the hole in the bedroom and the one in the living room, and attach the cable at the connector to the wire hook. Pull the cable through the hole and into the bedroom. Clean any debris from the connector by blowing on it.
6. Take the living room wall-plate cable connector into the bedroom and attach the cable F connector to the wall-plate outlet. Attach the wall plate to the wall with the screws from the living room as shown in *Figure 4-7d*.
7. Use a blank wall plate to cover the hole in the living room as shown in *Figure 4-7d*. Blank wall plates can be purchased at most hardware and home improvement centers.

Extending the Relocated Outlet to Add a New One

Suppose the decision had been to leave the outlet in the living room and to *add* a second outlet in the bedroom. The same steps of 1 through 4 as described for the relocation are followed, then proceed to step 8.

8. To the F connector on the cable through the hole in the living room wall, connect a two-way splitter. This cable is connected to the F-type *female* input connection of the two-way splitter.
9. Attach two short cables with male F connectors to the two F-type female connections remaining on the two-way splitter.

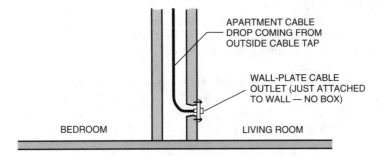

a. Present Cable Installation

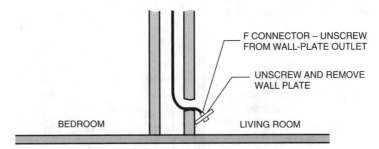

b. Remove Present Wall-Plate Outlet

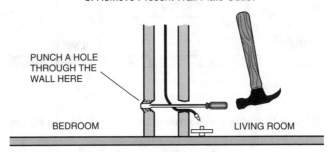

c. Punch Hole Through to Bedroom

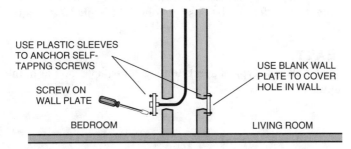

d. Finish the Relocation

Figure 4-7. Relocating a Cable Outlet from a Living Room to a Bedroom.

10. Push one of the two cables just connected to the two female F connectors on the two-way splitter through the bedroom hole. If you are unable to push the cable through, follow the instructions of step 5.
11. Connect the remaining cable to the wall-plate cable outlet that was removed in the living room.
12. Push the two-way splitter and cables through the hole in the living room wall. The hole may have to be enlarged, but be careful that the hole is not made so large that it can't be covered by the wall-plate outlet. Attach the wall-plate outlet back on the living room wall with the screws that were removed initially.
13. Go into the bedroom. Attach the cable with the male F connector coming through the hole in the wall to a new wall-plate outlet, and affix the wall-plate outlet to the bedroom wall. This is shown in *Figure 4-8*.

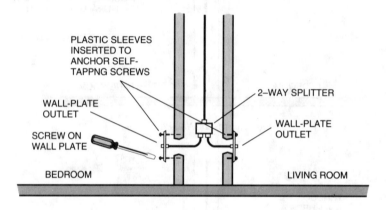

Figure 4-8. Adding an Outlet in the Bedroom from One in Living Room.

If you wish to construct your own cables by buying separate bulk cable and separate F connectors, refer to the instructions in Chapter 7 for attaching the connectors.

WHAT'S NEXT?

Now that we have looked at typical apartment systems, let's look at home systems in the next chapter.

Chapter 5
Home Installations

THE HOMEOWNER IS KING

If you own your own home, you have many more options for video than your friends who live in apartments. Typically, you have a yard in which you can install an outside dish if this is the kind of installation you would like. You also have the "privilege" of running cable throughout your home and adding up to four TV outlets per hardline cable drop.

Figure 5-1a shows a typical hardline cable installation in a home. The cable system signal comes from a cable company tap through a cable to the ground block mounted on the outside of the house. The ground block, shown in *Figure 5-1b* is an important device. It grounds the system and protects against power surges and static build up. It is also the starting point where homeowners can make changes to their cable TV services. A ground wire connects the ground block to a ground rod or to a power line conduit to provide a substantial electrical ground. From the ground block, a coaxial cable usually is run into the attic. The cable routed into the attic may be connected to a splitter. A splitter does just what the name implies — it splits the signal into several paths. The number of splits is dependent on the number of TVs you want to feed. For example, for two TVs, it would be a two-way splitter; for three TVs, it would be a three-way splitter. The distribution of the cable signal to two TVs through a two-way splitter is shown in *Figure 5-1*.

Normally, there may be only one TV, indicated by the solid lines in *Figure 5-1*. If the installation is for two TVs, the additional cabling is as shown in the dotted lines of *Figure 5-1*. Sometimes the splitter is mounted near the ground block, and sometimes the cable in the attic is cut and the splitter inserted.

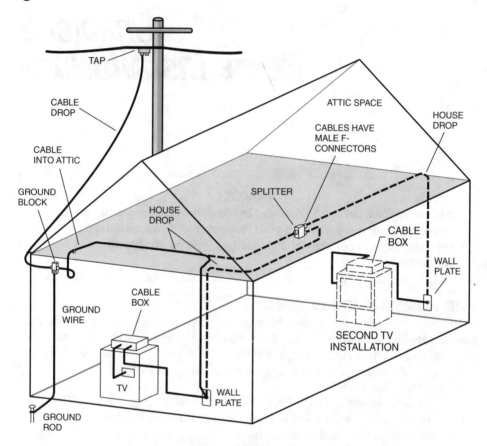

a. Home Installation

b. Ground Block

Figure 5-1. A Basic Hardline Cable Installation for One TV is Shown in Solid Lines, While a Second TV Installation is Shown in Dotted Lines. The Cable Drop Connects to a Ground Block.

Distribution Through a Splitter

A two-way splitter is shown in *Figure 5-2a*. The single input on one side receives the cable from the ground block (the incoming signal), and the two outputs distribute the signal to cable converters for the TVs. An easy way to accommodate up to four TVs (and/or FM stereo receivers) is to install a four-way splitter (shown in *Figure 5-2b*) in place of the two-way splitter and run the necessary cables to the other sets.

Typical Cable Input Into the Home

The same wall-plate outlet *(see Figure 4-2)* and F connectors *(see Figure 4-3)* shown in Chapter 4 are also used for home cable installation.

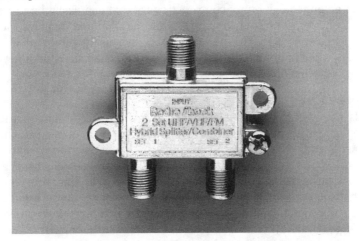

a. 2-Way Splitter

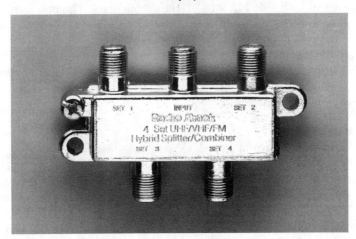

b. 4-Way Splitter

Figure 5-2. A Splitter Accepts a Signal Input and Distributes It into Separate Outputs.

HOME SATELLITE SYSTEMS

The home satellite system has the greatest variety of programming of all the TV video systems available. A typical installation is shown in *Figure 5-3*. The satellite dish must be located in a spot where there is an unobstructed view to the South. Some installations will have the satellite dish close to the ground because the view is clear. In other installations, the satellite antenna is raised on a pole to obtain a view free of obstructions.

Antenna Positioning

The antenna must be positioned to point at a particular satellite. There are several satellites available with different programming. Most modern satellite receivers have built-in antenna positioning control circuitry that can easily be programmed by the user. The receiver controls a motor on the dish and displays the current satellite on the TV screen. You can move the dish using the receiver's remote control. Older satellite antenna installations used mechanical cranks for positioning and did not have motors at all!

The Signal Path

The signals received by the satellite dish are focused on the antenna feedhorn. There is an amplifier in the feedhorn to boost the very weak satellite signal. Circuitry also installed in the feedhorn converts the signals to lower frequencies (usually from 950 to 1450 MHz) that are easier to process electronically. This signal is sent through the underground coax cables that run to the house and inside the house to the satellite receiver. Modern receivers have a circuit built in to descramble the coded signals. A software program for descrambling is necessary for the receiver. Older systems had separate, external descramblers. The receiver is capable of receiving either C-band frequencies (3.7 - 4.2 GHz) only, or a combination of C-band and Ku-band (11.7 - 12.2 GHz) frequencies. The underground cable bundle between satellite dish and receiver also carry signals for positioning the antenna and for providing power to the antenna circuitry.

If splitters or filters must be installed, they must be designed especially for the 950 - 1450 MHz converted frequencies. Conventional cable splitters and inline filters may not be used. This is important to remember in case modifications or changes are made to the system!

Local TV Stations

Most satellite systems have standard TV antennas installed to receive any local "off-the-air" TV stations in the area. The normal signal coupling from the antenna to the TV receiver (a typical example is shown in *Figure 5-4*) is either with 300-ohm "twin-lead" or 75-ohm coaxial cable. If a 300-ohm twin-lead is used, it is terminated in a VHF-UHF band splitter that has a 300-ohm input and VHF and UHF outputs. If a 75-ohm coaxial cable is used, it is again terminated in a VHF/UHF splitter, but it has a 75-ohm input. In either case, from the splitter the VHF connection is through a 75-ohm coax cable to the 75-ohm VHF connector on the TV, and the UHF connection is through 300-ohm twin-lead to the UHF terminals on the TV.

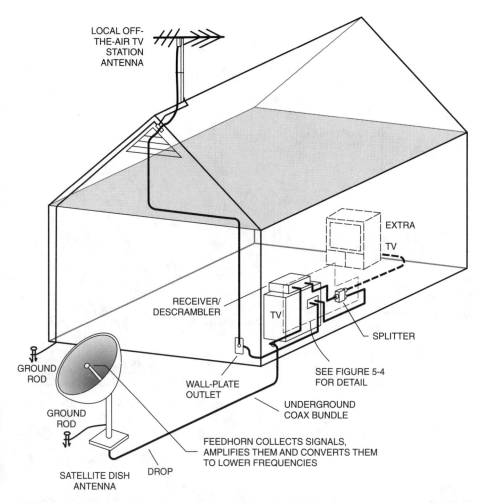

LOCAL OFF-
THE-AIR TV
STATION
ANTENNA

EXTRA

TV

RECEIVER/
DESCRAMBLER

TV

SPLITTER

GROUND
ROD

SEE FIGURE 5-4
FOR DETAIL

WALL-PLATE
OUTLET

GROUND
ROD

UNDERGROUND
COAX BUNDLE

FEEDHORN COLLECTS SIGNALS,
AMPLIFIES THEM AND CONVERTS THEM
TO LOWER FREQUENCIES

SATELLITE DISH   DROP
ANTENNA

Figure 5-3. The Drop That Runs from the Satellite Dish to Your House Has Multiple Cables Bound Together to Power the Feedhorn and Carry Signals to the Receiver/Descrambler.

The connection shown in *Figure 5-4* is a bit different, and many installations will need to be made this way. *Figure 5-4* shows the details of switching between the satellite receiver signals and the off-the-air local channel signals. The satellite receiver output (on Channel 3 or 4 VHF) is coupled to a high-isolation A/B switch as one input. The off-the-air local antenna signal is coupled to the other input. The output of the A/B switch is coupled to a VHF/UHF splitter, and the two outputs of the splitter are connected to the respective VHF and UHF terminals on the VCR, and the VHF/UNF outputs on the VCR are connected to the respective inputs on the TV. To provide additional flexibility, a splitter is inserted between the satellite receiver and the A/B switch so that a second TV may be added. Note also that a 300-to-75-ohm transformer is added right at the local TV antenna so all coupling is made through 75-ohm cable. Cable ready connections are also shown in *Figure 5-4*.

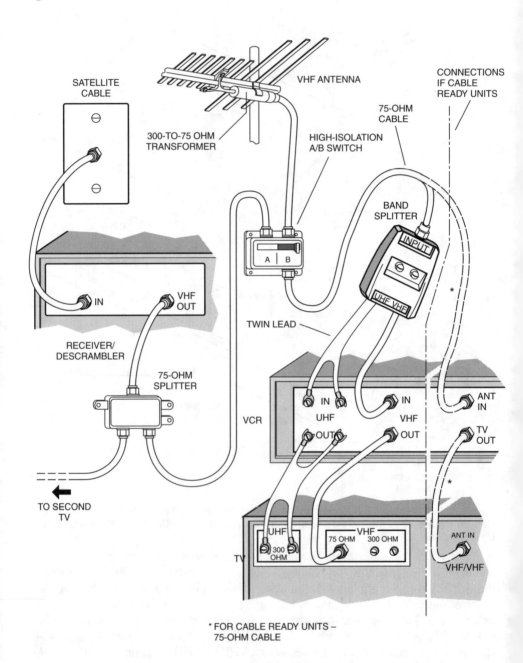

SATELLITE CABLE

CONNECTIONS IF CABLE READY UNITS

VHF ANTENNA

300-TO-75 OHM TRANSFORMER

75-OHM CABLE

HIGH-ISOLATION A/B SWITCH

BAND SPLITTER

INPUT

UHF VHF

A | B

IN

VHF OUT

RECEIVER/ DESCRAMBLER

75-OHM SPLITTER

TWIN LEAD

VCR

IN UHF

IN VHF

ANT IN

OUT

OUT

TV OUT

TO SECOND TV

UHF

300 OHM

VHF

75 OHM   300 OHM

ANT IN

TV

VHF/VHF

* FOR CABLE READY UNITS – 75-OHM CABLE

Figure 5-4. An A/B Switch is Placed Between the Satellite Receiver Output and an Off-the-Air TV Antenna So the TV Can Be Switched Between the Two. TVs/VCRs May Have a Separate Input, as Shown, or a Combined Input for Cable Ready Units.

WIRELESS CABLE SYSTEM (MMDS)
Antenna Positioning and Signal Path

The microwave antenna for the wireless cable TV system is usually mounted on a tripod on the roof to provide a good line-of-sight path to the transmitting station. The antenna is about the size of a newspaper. A TV antenna can be mounted on the same tripod to receive any local off-the-air stations, just as is done with the satellite system. A 300-to-75-ohm transformer is placed at the local TV antenna to couple the signal over 75-ohm cable to a diplexer on the microwave antenna tripod. The diplexer combines the signal from the two antennas so that only one cable is needed to couple the signals, through a ground block, to the MMDS converter box in the house. A special splitter is required at the converter box when a diplexer is used to input the MMDS signal to the "cable" input and the off-the-air signal to the "antenna" input. The output of the MMDS converter box is a VHF signal (channel 2 or 3), and is connected to the 75-ohm VHF connector on the TV. *Figure 5-5* shows a typical system interconnection. The cable signal input to the house is through a wall-plate outlet. *Figure 5-5* also shows how a splitter can be inserted at the output of the converter box to add a second TV.

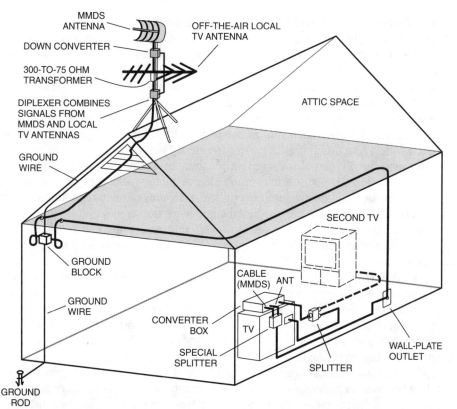

Figure 5-5. MMDS Microwave Signals and Off-the-Air Local TV Signals Are Combined in This Wireless Cable TV System.

Dual Purpose for the Cable Drop

On the wireless cable TV system, the house drop, the cable that brings a signal into the home, serves a dual purpose. Power is supplied through the house drop to the MMDS antenna. The power is needed by a down converter at the MMDS antenna. The down converter changes the frequency of the MMDS microwave signal (2.5 - 2.686 GHz) to a much lower frequency (222 - 408 MHz) containing the TV signal. The lower-frequency signal is sent to a converter box at the TV where the signal is unscrambled for TV viewing.

DIGITAL SATELLITE SYSTEM (DSS)

Antenna Positioning

When the Digital Satellite System antenna is installed, there must be a clear view to the South. Since it is an 18-inch parabolic dish rather than a 6- to 9-foot dish, it is usually installed on the roof at the edge of an eve. However, for security reasons, it may be installed on the chimney. Some installations are easy because there are no tall buildings or tall trees obstructing the southern view. Others can be quite difficult because many obstructions may be in the way. Sometimes trees must be topped and kept trimmed to make a clear signal path.

This Time The Signal Path — Ku-Band

A typical system installation is shown in *Figure 5-6*. There is a single coaxial cable from the DSS antenna to the DSS receiver. It comes to the house at the ground block and then to the receiver through a wall-plate outlet. The frequency of the signal collected by the DSS antenna is in the Ku-band (12.2 - 12.7 GHz). It is a much higher frequency than the C-band for the 6- to 9-foot satellite dish. The size of the antenna is much smaller because the wavelength of the signal is much shorter. Because the signal is digital, the bandwidth of the signal to be transmitted is much wider, so a much higher carrier frequency is required.

Because the TV signal is digital rather than an analog signal like all local off-the-air TV stations, the result is a much clearer picture and compact disc-quality sound. Some say the improvement over regular TV and satellite reception is fantastic! There is another advantage—digital signals are not as susceptible to electrical interference as analog signals are. However, rain and snow can block out DSS signals. The particles are near the wavelength of the microwave frequency and can absorb energy from the signal!

Like the MMDS system, the cable from the receiver to the antenna serves a dual purpose. Not only does it provide a transmission path for the signal, it supplies voltage to the DSS LNB (Low Noise Block downconverter) that translates the signal frequencies from 12.2 - 12.7 GHz down to 950 - 1450 MHz. For this reason, no ordinary[1] splitter may be in the coax line from the DSS antenna to the receiver. Splitters can interrupt the flow of voltage to the downconverter.

The interconnection of the DSS signal to the TV is as shown in *Figure 5-7*. The satellite receiver of *Figure 5-4* is replaced with the DSS receiver. There is no need for an A/B switch because the DSS receiver has one built in. The local off-the-air antenna is connected to the DSS receiver.

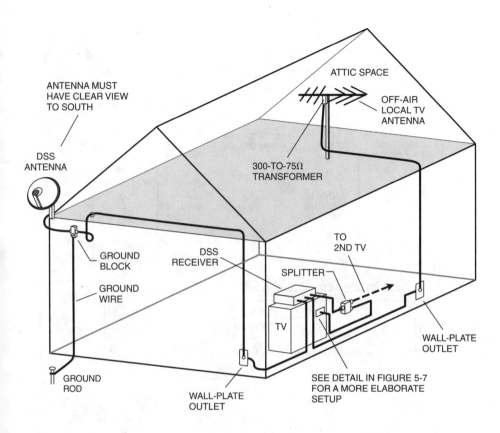

ATTIC SPACE

ANTENNA MUST
HAVE CLEAR VIEW
TO SOUTH

OFF-AIR
LOCAL TV
ANTENNA

DSS
ANTENNA

300-TO-75Ω
TRANSFORMER

TO
2ND TV

GROUND
BLOCK

DSS
RECEIVER

SPLITTER

GROUND
WIRE

TV

WALL-PLATE
OUTLET

GROUND
ROD

WALL-PLATE
OUTLET

SEE DETAIL IN FIGURE 5-7
FOR A MORE ELABORATE
SETUP

Figure 5-6. When the Local TV Station Antenna is in the Attic, the DSS Dish Will Be the Only TV Antenna Seen from Outside of the House

Local TV Channels

For local channel TV stations an off-the-air TV antenna must accompany the DSS installation. It may be in the attic, as shown in *Figure 5-6,* so it may not be seen from the outside. This is desirable if you have deed restrictions or ordinances against outdoor TV antennas.

Accessories

The receiver has a built-in field-strength meter used for antenna positioning. Once the antenna is manually positioned to receive a signal, the field-strength meter is used to 'tweak' the antenna to the maximum signal position. Perhaps, in time, your antenna may get slightly out of alignment due to weather or settling of your house. The field-strength meter can be used to optimize the position of the dish. If the DSS receiver is not set up to have signal response of 90 to 100 on the signal meter (indicated on the TV screen) during the antenna positioning, the weather effects will be much more pronounced.

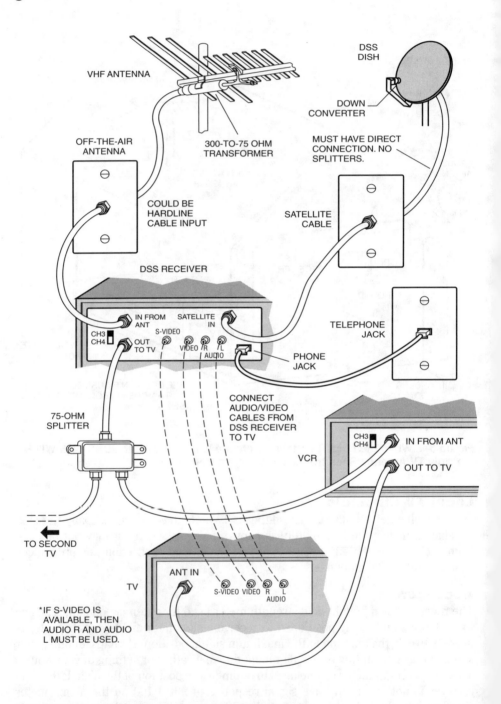

Figure 5-7. Combining DSS and Off-the-Air TV Signals.

Some receivers have twin outputs for multiple satellite receiver and TV expansion, gold-plated RCA-type input and output connectors, wide-band data port, and an S-video output. All the receivers can be controlled remotely, often with a universal remote.

There are many different programs offered by the two providers, DirecTV™ and USSB™. To use the system, you must subscribe to a programming service. The rates are competitive with hard-line cable.

The system must have access to a telephone jack if any of the "pay-per-view" channels are to be received. There is an access charge (about $2.00) for each telephone call. Recognition and billing is through the phone line.

All receivers, TV converter boxes, and descramblers need 120VAC (household line voltage) to operate.

HAVING A CABLE SYSTEM INSTALLED

Before you call the cable company to order service, stop and think about where you would like your TVs to be and where you might move them in the future. Hopefully, this will determine the ideal position of the cable wall-plate outlets and confine the area for each TV to about six to eight feet on either side of the outlet. Ask the cable installer to leave you extra cable for each TV. This will give you the flexibility to move the TVs around a bit.

What to Expect

Here is what you should expect when the cable installer comes to install your cable service. He will decide where the ground block is to be mounted, and how to run the cable drop from the tap to the ground block. After the outlet position is established, the installer will drill a hole in the wall behind the TV for the wall-plate cable connector outlet. If space is available and accessible, he will run the cable from the ground block through the attic and inside the wall to the hole for the wall-plate outlet. If he cannot get the cable down the wall, he will have to run the cable through an outside wall. From the ground block, he will have to tack the cable around the outside of the house to the point where it goes through the outside wall. Remember, each TV that is hooked up to cable will have a separate cable running to it. He may insert a special tube in the hole through the outside wall, or just push the cable through the drilled hole. In either case, he should weatherproof around the tube and/or cable so insects and moisture will not penetrate into the wall or your home.

While the installer is there, it is a good time to have him install any additional outlets you may feel you need, or you may install them yourself at a later time. If you plan to add outlets yourself, pay close attention to how the installer goes about his work.

Installing Your Own Outlets

Installing your own outlets is certainly possible. Complete instructions on how to do it yourself are presented in Chapter 7. Materials, tools, and step-by-step instructions are included. You may want to read them over before you call out the installer initially so you can more intelligently instruct him what to do. It may even save you some money!

Chapter 6
How to Recognize and Overcome Problems

TV PROBLEMS — EXPECT THE UNEXPECTED

It may be trite, but it's true. Into every life a little rain must fall. You've sunk a small fortune and lots of hard work into your video system, and on the day of the "big game" the cable goes out. Or there's a heavy snowstorm and DSS satellite reception goes to pot. Or those cute bushy squirrels that feast on your bird seed decide to feast on the cable running from your giant off-the-air antenna. No one particular system is safe from trouble. Problems can be the result of atmospheric conditions, animals, or hardware failure. Take heart. This chapter will help you avoid panic and be able to analyze what kind of problem you are encountering. Once identified, you can take the necessary action to restore everything to normal.

MAJOR NATURAL CAUSES
Some Areas are Not Favorable to Certain Systems

There are regions around the country that are not "friendly" to certain types of video systems. Regions of the country that are densely forested are not well suited for wireless cable or satellite dishes. If trees block the view of your satellite or MMDS dish, you will likely have poor reception. The density of the foliage on the trees will reduce the amount of incoming signals. As a result, foliage growth during spring and summer will make the problem worse, but come fall, when the leaves drop, reception may improve dramatically! Hardline cable TV is the preferred video system in these regions.

Other natural barriers that can cause trouble are mountains and hills. Most cities and housing developments are built in valleys and open land *below* the mountains and hills. In these areas, wireless cable TV is rarely available. If you have an unobstructed view to the South, a satellite system will work just fine. If there is a tall hill to the South, it will block large-dish and DSS satellite signals. Even off-the-air outdoor antennas might just be pretty aluminum sculptures if you're stuck in the lowlands. Here again, hardline cable will be your only alternative.

Man-Made Objects Also Block Signals

Man-made objects will block signals, too. In the areas where wireless cable TV is available, its transmitter is usually located on top of a tall building or tower. Ironically, other tall buildings, large water towers, and huge power line structures will obstruct the wireless microwave signal if they are in the path between your home and the transmitter.

Why do forests, mountains, and structures block TV signals? The frequency bands that satellite and MMDS systems use are high up in the electromagnetic spectrum. These higher frequencies are basically line-of-sight. They do not penetrate solid objects at all. In fact, they reflect off of solid objects.

Lightning and Rain

Major weather conditions that affect TV video systems are rain and lightning. Lightning strikes pose a threat to wireless cable TV and tower-mounted off-the-air antennas. As you recall, a wireless cable TV antenna is mounted on top of the subscriber's home; a mast-mounted antenna may indeed be the highest structure in the area. The antenna and the mast make good targets for lightning. It is essential that the antenna and mast be electrically grounded. If you live in an area with frequent thunderstorms, make sure that you or the installer ground the system and take all the steps necessary to reduce the likelihood of damage from a lightning strike.

Home satellite and Digital Satellite systems are less likely to get struck by lightning because of their position (close to the ground or near the home). Nevertheless, it is important to remember to properly ground these systems as well. Grounding techniques are shown in the home installations in Chapter 5.

Hardline cable TV can also be damaged by lightning. In areas where the cable feeder lines are on telephone poles, power surges are a possibility. Lightning can strike a main feeder line on a telephone pole, which in turn sends a surge throughout the system. Because of this, a ground block is installed on every home between the cable from the company and the wiring in your home. As we have shown previously (see *Figure 5-16),* a ground block is the device used to couple the exterior cable from an antenna or a cable system to the cable inside the house. In addition, the ground block grounds the entire setup through a ground wire connected to a ground rod.

Microwave signals are also scattered by moisture. You'll notice that on a clear day, most TV reception (off-the-air, MMDS, and satellite) is perfect. On a rainy or snowy day, high-frequency MMDS and satellite reception might get quite fuzzy. Even hardline cable TV reception might be fuzzy, as the cable company relies upon satellite dishes and off-the-air antennas to receive all of its programming.

Rain Damage

As you read through this section you may have the feeling that hardline cable TV is pretty much trouble-free, except for lightning. It isn't. Its primary enemy is rain. When there is heavy rain or prolonged rainfall, water can accumulate in the cable lines. A considerable amount of moisture is collected within feeder lines and regular coaxial cable. This is usually due to poorly attached F connectors at various spots along the cable. If your cable regularly suffers a blackout after a storm, you can be certain that an invasion of water somewhere in the cable system is the culprit.

Hail and High Winds

Common sense tells you that hail and high winds are hazardous to people. They're also hazardous to video installations. A bad hail storm can render a home satellite system useless, especially if the dish is made from wire mesh. Large hail can significantly alter the contour of the dish so that it no longer reflects and focuses incoming signals properly. Once the shape of the dish is battered out of shape, it becomes an interesting yard sculpture. The feedhorn is also very vulnerable. If the feedhorn, which contains major amplifying and down-converting circuits, takes a direct hit from a large hail stone, it can be destroyed! Hail is usually accompanied by high winds. High winds may also affect home satellite systems. If the satellite dish is not anchored properly, high winds will knock it out of alignment.

Wireless cable TV systems can be damaged by hail and high winds. Recall that the wireless microwave antenna is metal. It can be dented and broken by a severe hail storm and, if not properly anchored, can be toppled by steady high winds. The mast should be well guyed and the dish well secured to it. Since this system is installed by a professional, he will most likely ensure that it is secure against high winds. If a severe storm hits, check the system to make sure it is still secure.

New Systems

There's good news if you have recently purchased or are interested in purchasing a satellite system! Experience has been a good teacher for manufacturers of satellite systems. The antenna dishes are now more ruggedly constructed. A good example is DSS. The 18-inch satellite dish is made of a highly durable plastic (this same material is now being used to make the larger dishes) that can withstand some very rough weather. The only serious hail threat is a direct hit to the feedhorn or arm holding it to the dish.

Snow and Ice Build Up

Snow is pretty to look at and fun to play in, but it can be a problem for large satellite dishes. Heavy snow fall can build up inside a satellite dish and attenuate the incoming signal. The result is poor quality pictures. It is easily corrected, though by carefully removing the snow from inside the satellite dish. A broom should do the trick, but be very careful that the position of the feedhorn is not changed. If a large accumulation of ice forms on the dish, you have to be careful when removing it. Do not attempt to scrape or chip the ice from your satellite dish because you face the risk of damaging it. A better technique is to use a portable hair dryer. Slowly melt the ice and, with just a little help, it will easily slide off the dish.

Ice can also affect hardline and wireless cable TV. Chances are, ice will form on part of the cable hardware that is on the *cable company's side* of the ground block. If you suspect your hardline or wireless cable TV is affected by ice, call your cable operator and report the problem.

Animal Interference

It's true. Animals like cable TV, but not to watch. To eat! *Figure 6-1* shows a cable that an animal has chewed. Small mammals like to gnaw on coaxial cables. In fact, they find cable lines an especially tasty treat. Even ants have been known to disrupt cable service! They enter poorly sealed boxes and short out the electronic circuits inside. Animals often chew on the cables until the center conductor is exposed. Once the shielding and outer insulation are chewed off, two things happen. First water readily enters the cable and the associated problems result. Second, no shielding will result in a snowy picture.

You are receiving grainy pictures on your TV and notice evidence of visits by small animals. Chances are you've found the cause of your TV problem. Examine your attic and crawl space for a chewed cable. If you find one, our advice is to call an exterminator *before* you replace the damaged cable. If you do not eliminate the animals, they will happily eat their way through the new cable. Call your cable company to install new cable or replace it yourself. If you opt to do it yourself, refer to Chapter 7 for complete instructions.

Figure 6-1. Animals Have Chewed Through the Insulation on This Coaxial Cable.

Earthquakes

There are a few other natural events that are totally beyond your control; earthquakes are one of them. A small tremor can leave a home satellite dish a little off target. If you believe this to be the case, get out your level and compass and check your dish alignment, and the possibility that the mounting post is no longer exactly vertical. This, too, will cause a "noisy" picture because of the low signal strength. Use the level to determine if it has moved from perpendicular. Hopefully, the post mounting is still solid, if it is, readjust the antenna direction to reestablish the signal strength.

Solutions Follow Causes

No video system is foolproof. Problems can result from atmospheric conditions, animals and unexpected physical events. Remember, the first part to solving a problem is to identify its cause. With the cause in hand, it will help you discover a solution, or help you advise someone you have entrusted to fix your system.

Some Systems Are Insurable

Here is good news for home owners! If you are going to invest in a large-dish satellite system or DSS, call your insurance agent. He can add your system to your home owner's policy. The cost is nominal, and you'll find it worth the extra expense. Unless you've got money to burn, it makes good sense to get the additional coverage!

AVOIDING INSTALLATION PROBLEMS
Kinks in the Cable

A problem that may occur when running coaxial cable throughout your house is accidentally breaking the center conductor. If the center conductor of your coax cable is broken, your picture will be forever snowy. Cable is designed to withstand tugging along its length and F connectors hang on pretty tightly when properly crimped, but cable is not tolerant of kinking. An example of kinked cable is shown in *Figure 6-2*. It is possible to break the center conductor while the shielding and outside insulation remain intact, so a break in the center conductor can go unnoticed. When looping coax, keep the diameter of the loop no less than 8″. Do not twist or tangle the cable when pulling it down inside a wall or through an attic. Do not jerk the cable. In other words, handle it with care and don't force or stress the cable. Let it move freely as it is installed.

Figure 6-2. Avoid Making a Kink in Your Cable. This Section Will Have to be Replaced.

Be Careful with Staples

Another thing that can cause a snowy picture is a staple gun. If you plan on using staples to keep coax cable in place, do it with care. *Without* a lot of practice and the right staples, it is easy to damage the cable or staple through the center of the coax cable. Look at the result of a misaimed staple gun in *Figure 6-3*. The cable shown was stapled to the trim around a soffit. A staple penetrating the insulation and shield can cause a short circuit between the shield and center conductor. One staple piercing the cable, even though it doesn't produce a short, can cause a snowy picture. A few penetrating staples, and you may not have any picture at all! If you must use a staple gun, be sure to practice up a bit on some scrap cable before tackling the real thing.

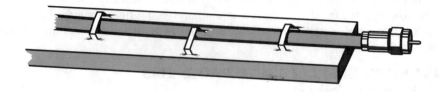

Figure 6-3. One Staple Through the Heart of Your Cable Can Short All Signals. Result — No Picture!

There is a very good alternative to staples. It is shown if *Figure 6-4*. It is a nail-in coax clip. The clips are easy to use and provide a neat and tidy installation. Install them every 12 to 15 inches along the cable run. They are readily available in most electronics parts stores and come in white and black.

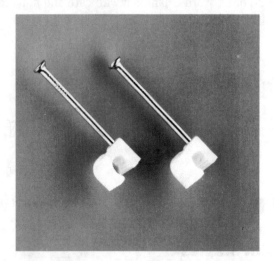

Figure 6-4. Nail-In Coax Cable Clips for Securing Coax Cable.

Use New Coax Cable

Start fresh with new coax cable when wiring your home. This is a rule that should never be violated. Never use old or left-over coaxial cable to make jumpers. Don't be tempted to use the stuff your brother-in-law found in his garage! Unless you know for a fact the cable is good, *don't use it*. Visit your local Radio Shack store for fresh, UL-approved coaxial cable.

F-Connector Woes

If not properly installed, F connectors can cause a snowy picture, an intermittent picture, or no picture at all. If you suspect this to be your problem, check for these conspicuous signs. Look inside the F connector. See if the aluminum braid is touching the center conductor in any way. This will cause a snowy picture. To correct this problem, attach a new F connector to the cable. Check if the connector is screwed on tightly (if the screw-on type) or pushed on all the way and seated tightly (if the push-on type). Make sure all the F connectors installed on your video equipment and accessories are hand tight. If not, intermittent pictures can result. When you look inside the F connector, make sure the center conductor is not smashed into the outside housing. Screwing on the connector without properly inserting the center conductor into the center hole can cause the center conductor to be bent and jammed inside the connector. No connection means no picture.

Check for connectors that have not been installed properly. *Figure 6-5* shows an example of a connector that was installed and the shielding was not cut properly. It is extending outside the connector, and the connector is loose on the cable. You should not be able to grab the cable and pull it out of an F connector. Improperly installed F connectors may cause lines to appear on your television. They may also cause your picture to jump or have "sparklies" when jostled. If a cable outside your home has a loose F connector, the cable will flex in the wind. This can cause problems to your own and your neighbor's reception. Chapter 7 has detailed instructions on how to properly install F connectors.

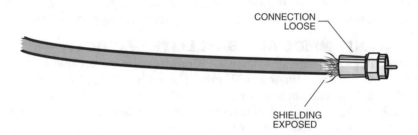

CONNECTION
LOOSE

SHIELDING
EXPOSED

Figure 6-5. This Poorly Installed F Connector is Easy to Spot.

CABLE COMPANY REPAIR SERVICES

If a problem develops with your hardline cable TV system and you are sure that it isn't caused by any of your hookups or equipment, call the Customer Service Department of your cable company. After all, part of your monthly cable fees go to maintain the equipment. If it is a cable company problem, bad picture quality, or the picture is out completely, chances are some minor problem has occurred in the cable company's distribution network. A repairman has probably already been dispatched to fix it.

A very common problem for interactive cable systems is that the power to the cable decoder box has been interrupted. If this occurs, or if your cable decoder box has developed another problem, it will have to be reset or replaced. Make the call to the customer service people. Very often it can be tested remotely by your customer service representative or their on-call technician, and the problem can be corrected over the telephone.

Another common problem to be handled by the cable company customer service representative is when a pay-per-view movie that has been ordered is not received. Such complaints, billing issues, and technical problems are specific reasons why the customer service organization has been set up. Not only is there a customer service person and an on-call technician, but there is a supervisor and general manager available in case complaints need to be escalated to higher levels. The customer service representative answers the phone and solves as many problems as he or she can; the on-call technician solves the technical problems; the supervisor oversees the customer service staff; and the general manager is responsible for the smooth operation of the entire cable company.

While a disruption to your cable service may be unpleasant, please keep in mind that all mechanical and electrical devices are subject to failure. There is no need to get upset with your cable provider. It is in their best interest to keep you satisfied so you will continue to pay for their service. With the increased competition from Direct Digital Satellite, you can be certain that your local cable company will do everything reasonable to keep you happy. If you do get into a position where you feel like you are getting the run-around on a complaint, be patient. Don't get discouraged. Stick to your guns. If necessary, find neighbors and friends who have similar problems and band together to bring more attention to your problem.

WHAT ABOUT MMDS AND SATELLITE SERVICE?

If you have a large-dish satellite system, DSS, or MMDS wireless cable system and have a problem, are you on your own? Hardly! The sellers of the equipment and the manufacturers have customer service organizations, just like hardline cable TV providers. If you lease equipment, the service provider is responsible for repairs. MMDS providers are responsible for all of the hardware and coax cable hookups, just as hardline cable providers are. Use the tips in this book to isolate a problem in your system; then give them a call. In the case of a satellite system, maybe there is technical difficulty with the satellite. Perhaps there is trouble with their uplink station equipment. Before you start to panic and think your hardware has failed, locate a customer service organization through the seller, distributor, or manufacturer, make a call, and put your mind at ease.

SUMMARY

In this chapter, we have discussed problems that occur due to natural causes and installation. Problems that are easily recognized due to the physical appearance of the components. Chapter 9 discusses troubleshooting techniques for all types of problems.

Chapter 7
Video Installations –
The Right Way

DO IT YOURSELF

If you are handy, and you'd like to save some money, you can pre-wire a home under construction or make changes to existing video wiring. This chapter is all about finding the right tools and right materials, and the right way to put everything together. Follow these instructions, and, with some practice, you should be able to duplicate what your cable company installer would do. What's more, these guidelines will be invaluable for any type of video project you choose to tackle.

Why go through this trouble? One of your authors has been a service technician for more than 12 years. Many of his service calls have been due to poor quality installations; more specifically, improper fittings and poor-quality cable were installed by a "do-it-yourselfer." For this reason, this book provides instructions, that if you follow them, you should have no need to call a professional installer to fix up a botched job. The installations will be quality installations from the start.

GETTING TO KNOW THE EQUIPMENT

Here is a list of the more commonly-used equipment for making a new home installation, or upgrading an existing one. The names and brief descriptions will help any do-it-yourselfer understand the function of each component.

A/B Switch — A switch that allows you to select between two or more inputs, e.g., a switch to select between cable and outdoor antenna for TV signals. If the switch is used for video signals, the switch should be a high-isolation type.

Coupler or Barrel Connector — (F-81) A threaded cylinder-shaped device used to join two male F connectors together.

Coaxial Cable — A multilayer cable used to distribute TV signals. This cable consists of a copper core surrounded by an insulating layer of polyethylene. It is wrapped by a thin aluminum foil, over-wrapped with one or more layers of metal braid, then covered with a protective external vinyl coating. It comes in five types: RG-6, RG-8, RG-11, RG-58, and RG-59. Each is external available as single, double, triple, or quadruple shield types. RG-8 and RG-58 are 50 ohms impedance; the others are 75 ohms. RG-6 is the most popular type you will come across with hardline cable systems or home antenna systems.

Hardline Cable or Video Convertor (Decoder) — A device used to enable a non-cable-ready TV to receive a cable signal. It is used to convert scrambled cable signals into usable pictures for your TV. It is needed for premium channels and pay-per-view events. It serves as the cable channel selector.

F Connector — Metal connector at the end of a coaxial cable. F connectors are available in three different sizes to fit the different types of coaxial cables.

Ground Block — A device used to ground all the cable wiring in your house. It terminates the cable drop from the hardline cable company tap, and a ground wire runs from it to a good electrical ground.

Hex Crimper — The hand tool used to fasten F connectors to coaxial cable.

Matching Transformer — Enables any TV that has the "old fashioned" screw-type connectors for 300-ohm antenna twin lead to accept a 75-ohm cable input. The transformer connects between the antenna terminals on the back of the TV and the F connector on the cable. The transformer is also used at the 300-ohm antenna to match the antenna to a 75-ohm cable used as a lead-in for the TV signal.

Splitter — A device used to divide or "split" a single incoming cable signal to connect to two or more video devices. It is usually located in your attic. Splitters are available with two to five output ports, which distribute the video signal to from two to five sets.

QUALITY PRODUCTS AND WHERE TO SHOP

You will need special tools and hardware (connectors, couplings, switches, splitters, etc.) to ensure a professional and safe installation. Don't cut corners when purchasing tools and equipment and you'll find your installation will go much smoother, hold up better over time, and provide a better quality picture than if you used inferior quality products.

When planning your TV video system, the first decision is probably coaxial cable. You need to determine what type and the amount of coaxial cable you will need for the job. We cannot emphasize enough the need to use good quality coaxial cable. There are brands of coaxial cable that use copper braid shielding only with no aluminum foil wrap. This is not acceptable for the type of wiring you will be doing.

Always buy coaxial cable with aluminum foil wrap shielding and aluminum or copper braid shielding. Choosing the right coax will ensure the best picture quality, will not leak spurious signals that can create interference to aviation and law enforcement communications, and will last for many years without physical deterioration. *Figure 7-1* shows detailed views of a typical RG-6 coaxial cable, which is an excellent coax to use. If you feel you have severe electrical interference, use the RG-6QS because it has a quad shield.

Parts and Equipment List

Your authors highly recommend getting everything you need for your installation needs in order to complete the job just like a professional. To make your shopping easier, we've put together a list of tools and equipment that may be purchased at most electronics parts distributors, complete with descriptions.

TOOLS
Description

6" Diagonal Wire Cutters
"PRO" Crimping Tool for RG-59 and RG-6 F connectors
Precision Coax Cable Cutter
Coax Stripper. Adjustable blades for 3/16" to 5/16"
    diameter cable

CABLE (Bulk)
Description

RG-6 White 75-ohm Coax
RG-6 Black 75-ohm Coax
RG-6QS Quad-Shielded 75-ohm Coax
RG-59 75-ohm Coax
RG-8 50-ohm Coax
RG-58 50-ohm Coax

EQUIPMENT
Description

QS-56 For RG-6 QS quad-shield cable
Gold-Plated CF-59 (F Connector) for RG-59 cable
Gold-Plated CF-56 (F Connector) for RG-6 cable
Standard CF-59 for RG-59 cable (crimp)
Standard CF-56 for RG-6 cable (crimp)
F-81 coupler (barrel connector) joins two F connectors
Coax feed-through bushing for RG-59
Coax feed-through bushing for RG-6
Ground rod and clamp
Ground Block
Wall-Thru lead-in tube
Wall-Plate Outlet (white)
Wall-Plate Outlet (brown, wood grain)
Coax Nail-In Clips
Splitter (2-Way)
Splitter (4-Way)
High-Isolation A/B Switch

OTHER HOUSEHOLD TOOLS
Description

Electric or Cordless Hand Drill
Sharp Utility or Pocket Knife
Slip-Joint Pliers
Ruler or Measuring Tape
3/8" Wood Drill with 12" long Shaft
Flashlight

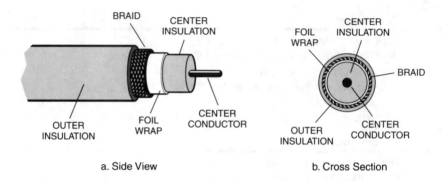

a. Side View   b. Cross Section

Figure 7-1. Two Views of Coaxial Cable Suitable for All Video Installations.

Supply Warehouse

The TV supply warehouses stock the top-quality equipment that professional cable installers use. These supply warehouses carry 2-way, 3-way, and 4-way splitters, F connectors, and the right types of coaxial cable.

Home Improvement Centers

A building materials retail store (such as Home Depot, Home Base, Builder's Square) is another place where you can find the tools and supplies you need for video installations. Most home improvement centers carry coaxial cable cutters, cordless drills, 3/8-inch wood drill bits with a 12″ long shaft, wall plates, and *fish* tapes.

PLANNING AN INSTALLATION

Do you know where you want your cable outlets? Think about where you would like the TV. Is it alone in a corner, or is it grouped together with an entertainment center? If this is a new house, look at your floor plan or blueprint. Make this decision *before* drywall and insulation are put in place.

Common Questions

If you are planning to install the cable yourself, whether pre-wiring a new house or moving or adding to an existing system, here are some questions you should be asking yourself:

1. **Is the wall accessible?** Locate the wall on which you want the cable outlet. It there is an attic or crawl space over or under that wall, then there should be no problem adding an outlet to that wall.
2. **Is this an outside wall?** If you are faced with no easy internal access, you will have to bring the input signal cable through an exterior wall. You will have to drill a small (3/8″ diameter) hole through the wall to bring the cable inside. If you drill a hole in your exterior wall, you must be sure to seal it. Use a "wall-thru" tube and seal it with silicone rubber to prevent air, water, or insects from coming into your house.

3. **Is the garage on the other side of the wall?** If there is a garage attached to the house, it may be possible to run a cable down an inside wall in the garage to a good access point into the house. Be careful; there may be insulation inside the wall that can slow you down. Work diligently and with the right tools, and you'll do it like an experienced installer! You may again want to use the wall-thru tube mentioned earlier for the access into the house.

4. **Can a cable be brought down inside a wall?** If there is an attic or crawl space over or under the wall, it probably can be done. You will have to drill a hole in the top plate or bottom base plate and push the cable down or up to insert it into the wall.

The preceding questions are designed to help you understand how an installer would approach a situation. Learning to think more like an installer will aid you in deciding what you want *before* you start a project. Knowing what you want beforehand will make the installation go much smoother.

Most Components and Techniques are Common to All Systems!

In this chapter, there are many references to hardline TV cable systems. Although this particular TV cable system is featured, all systems share most core elements. The procedure for adding an additional outlet to a hardline cable system is the same for adding an outlet to a wireless cable system, home satellite system, or Digital Satellite System. The only major difference in components is the type of converter box used. The uses of internal interconnecting coaxial cable, F connectors, splitters, and ground blocks, are common among the different TV video systems. If you master the use of these devices, you will be able to wire up *any type* of video system. However, be aware that the coaxial cables from the satellite dishes to their receivers may be special cables for transmission of the down-converter frequencies. If these cables are damaged, the special coax must be used as the replacement.

Exterior Wall Installations

Figure 7-2 shows how a hardline cable system is installed if the cables must be installed on the exterior walls of a home or apartment. The method is called "wrapping" or "tacking." Once the hardline cable TV outlet is located behind the TV, the cable drop goes down and through the exterior wall instead of into the attic and down into an interior wall. The wrapping or tacking gets its name because cable clips, which are plastic clips with small, thin nails through them, are used to attach the cable runs to the exterior wall(s). The cable runs from the ground block to where it goes through the exterior wall into the house. If two cable wall outlets are required, a splitter can be inserted in the cable run.

CAUTION

When drilling through walls make certain there are no electrical wires or water pipes in the way.

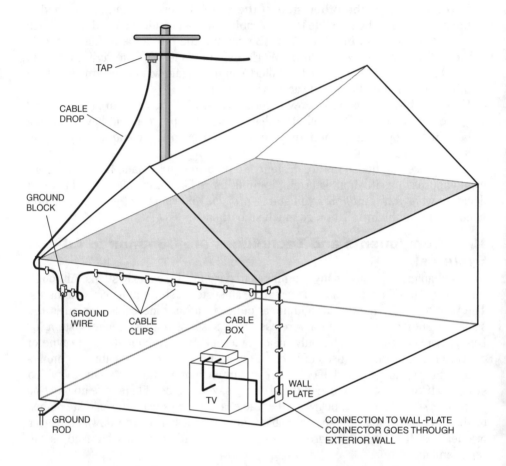

TAP

CABLE
DROP

GROUND
BLOCK

GROUND
WIRE

CABLE
CLIPS

CABLE
BOX

GROUND
ROD

TV

WALL
PLATE

CONNECTION TO WALL-PLATE
CONNECTOR GOES THROUGH
EXTERIOR WALL

Figure 7-2. When the TV is Located on an Exterior Wall and there is no Access for a Wall Drop, the Cable Needs to be Wrapped on the Outside of Your House.

BEGINNING AN INSTALLATION

The key to success with any cable hookup is to learn how to properly attach an F connector onto a cable to ensure good signal reception and a clear picture. To help, there are step-by-step instructions to explain how to prepare the coax cable and properly install an F connector.

Coaxial Cable Preparation

To prepare a coax cable for installing an F connector, and to do the best quality job, you will need the tools shown in *Figure 7-3a, b,* and *c.* Use the precision coax cable cutter of *Figure 7-3a* to cut the cable to length, and use the coax cable stripper of *Figure 7-3b* to strip the cable down for mounting an F connector. It will handle 3/16″ to 5/16″ diameter coax. Use the crimp tool of *Figure 7-3c* to crimp the F connector to the braid shield.

If you are experienced in soldering, preparing cables, and mounting connectors, then you can prepare the coax cable using only a sharp knife and a diagonal wire cutters shown in *Figure 7-3d.*

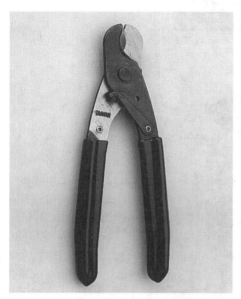

a. Precision Coaxial Cable Cutter

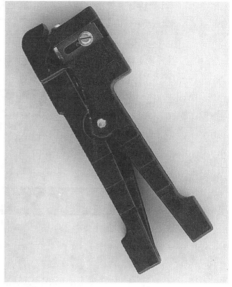

b. Coaxial Cable Stripper

c. Professional-Style Hex Crimping Tool

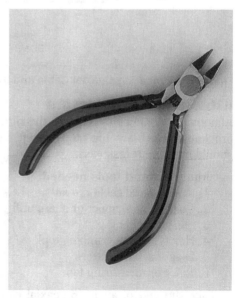

d. Diagonal Wire Cutters

Figure 7-3. Special Tools for Coaxial Cable Installation.

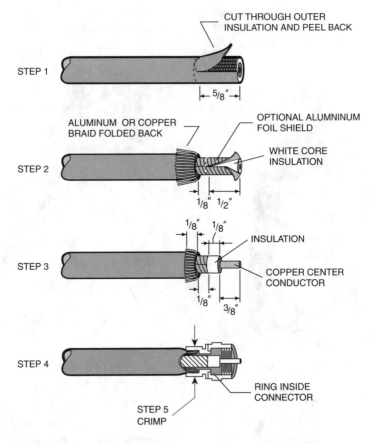

Figure 7-4. Preparing Coaxial Cable and Installing an F Connector.

Attaching F Connectors

Attaching F connectors requires the right materials, the right tools, and the right procedure. Here are the step-by-step instructions illustrated in *Figure 7-4,* that should make your task easy.

Materials and tools needed:
1. RG-6 coaxial cable (*do not* use RG-59)
2. Coax cable stripper or sharp utility or pocket knife
3. Slip-joint pliers
4. Hex-type coax crimping tool
5. Diagonal wire cutters
6. Ruler or measuring tape

1. Take the end of an RG-6 coaxial cable and measure back 5/8″ from the tip. Using the coax cable stripper or a knife, make a circular cut through the outer insulation, peel back and remove the insulation. Be very careful not to cut through or nick the aluminum or copper braid shield.

2. With a knife blade point, unbraid the aluminum or copper shield and fold it back over the outer insulation. Move back 1/2" and cut through the optional aluminum foil shield and peel it away. Do not cut through the white core insulation underneath.
3. Measure back 3/8" and with the coax stripper or a knife cut through the white core insulation to the center copper conductor. Make sure the aluminum or copper braid is folded back. Trim the braid to 1/8" overlap as shown. Make sure no braid wires are touching the center conductor because it will result in serious degradation of your picture.
4 . Slide the F connector over the white core insulation and the aluminum braid. Push down and twist the fitting until you see the front edge of the white core flush with the front edge of the ring inside the connector. Make sure no shielding is exposed out the rear end of the connector.
5. Now you will need a 0.360" hex shape crimping tool. Do not use *pliers!* Pliers will not work as a crimping tool. Fit the tool over the connector and crimp the connector onto the cable. Make sure it is on tight and will not pull off.

Checking Your Work

This is a quick check list of things to look for after you have put on a fitting:

Crimp Check: First, make sure you've made a good physical connection. After you crimp a fitting, give a gentle tug on the connector to make sure it will not pull off the cable.

Inside the F connector: Look into the open end of the connector. Make sure that the white core is flush with the copper ring inside the fitting. Make sure the center copper conductor is straight and positioned equally from all sides of the connector.

Braid: No braid should touch the center conductor. No braid should stick out the back end of the connector.

CRAWL SPACE INSTALLATION

A crawl space installation is relatively easy to do. The crawl space can either be the attic above a room or it can be the area below, such as a basement or the space under a house if the house has a pier and beam construction. First, find the entrance to the crawl space. It maybe stairs to the attic, an opening in the floor, a door to the basement, or an opening in the side of a house. Make sure the area that you want to work in is large enough for you to gain access and maneuver in.

Materials and tools needed:

1. Electric or cordless drill
2. 1/8" wood bit 6" long
3. 3/8" wood bit 12" long
4. Flashlight
5. Coat Hanger
6. Wall-Plate Outlets
7. Electrical Outlet Boxes
8. F Connectors
9. RG-6 Coaxial Cable (*do not* use RG-59)

Below the House

Here are the installation steps if you are installing a cable from a crawl space below the floor. Use *Figure 7-5* as reference.

1. Locate where you wish to install a video outlet. Note the room location and look for identifying wall layouts. Enter the crawl space and try to identify the location of the wall into which a cable is to be inserted. Note if there are electrical wires, conduits, or copper water pipes in the vicinity. You do not want to drill into these from above.

2. *Inside the house*, lift a baseboard away from the wall and drill a hole at an angle through the floor where the wall meets the floor as shown in *Figure 7-5a*.

3. After drilling the hole, drop a marker through it. A marker can be a coat hanger, *fish* tape, or a piece of electrical wire (at least 4 feet long). This makes it easy to locate the hole you drilled into the floor from above.

4. Enter the crawl space, locate the marker, locate the wall and pinpoint where the hole(s) are to be drilled to insert the cable. Remember that the wall is 3½" thick plus the thickness of the wall covering on each side (drywall is usually 3/4"). Remove the marker.

5. Drill a 3/8" hole for each cable through the floor and base plate and push a cable up through the hole.

6. Inside the house, fish the cable through a hole in the wall that you have made for the electrical outlet box to hold the wall-plate outlet. Attach an F connector to the end of the cable and connect it to the wall-plate outlet as shown in *Figure 7-5b*.

7. Go back to the crawl space and run the cable inserted in the wall across the floor supports and connect it to the cable coming from the ground block. There are actually two cables shown in *Figure 7-5*.

8. In *Figure 7-5b*, a splitter is shown to supply the two cables to the new outlets. To ensure a good picture, immobilize the splitter by mounting it to a wooden floor support by the crawl space entrance. Mount it high enough off the ground to avoid water damage.

9. Follow steps 1 through 7 for any additional outlets in other rooms. Run your lines to additional splitters before connecting the primary line to the ground block outside the house.

Attic Installation — No Fireblock

If you would like to install additional hardline cable TV outlets in your home when you have an attic crawl space and the walls have no fireblocks, here are the installation steps. Use *Figure 7-6* as reference.

1. While in the room at the chosen spot for your outlet, see if there is an air vent on the ceiling above it. In the attic, the distribution air ducts will be connected to the ceiling outlet. This will help you find the top of the wall you want. Once you are in the attic, you will find that locating the wall may be a little difficult because of insulation or flooring in the attic. Here are two techniques to get a little help:

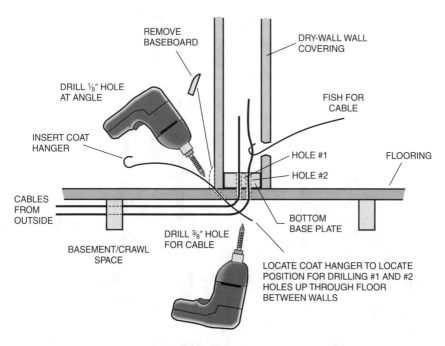

a. Locating, Drilling and Fishing Cable

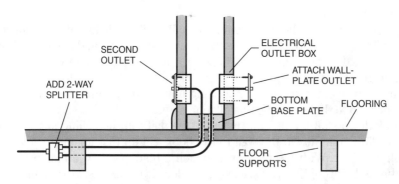

b. Finishing Touches – Installing the Connectors and Wall Plates

Figure 7-5. Installing Cable in a Crawl Space Below the House.

A. Take a small, thin 1″ nail and push it up through the ceiling just above the outlet position. This nail will mark the wall where you will drill through the top plate of the wall in the attic. In the attic, you will probably have to move insulation (either blown in or paper lined) to locate the nail.

B. When in the attic, have someone stand by the wall and tap on it until you locate the wall and the top plate.

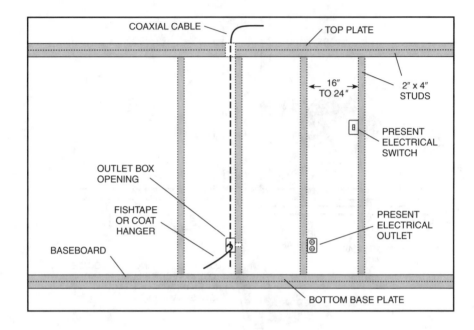

Figure 7-6. Installing Wall-Plate Outlet in Wall without Fireblock.

2. Once you have found the top plate of the wall, drill a hole in the top plate with a 3/8″ drill bit. The depth of the hole will be at least 1¾″. Next, drop down about 6′ to 10′ of coaxial cable (preferably RG-6) through the hole and into the wall. If another person is available, have them listen to make sure that the cable is dropping down completely, and not being bound up by a fire block. If there is a fire block, go to the next set of installation steps.

3. Go back down stairs. Find your marker (the nail) on the ceiling and imagine a straight line going down the wall to the floor. Go up that imaginary line about 13″ off the floor. With a screwdriver, punch a hole in the wall large enough to pull the cable through. If you are using an electrical outlet box to install the wall-plate outlet, cut a hole for the outlet box and mount the box.

4. Unfold and straighten out a wire coat hanger. Make a hook at one end of the wire large enough to grab the cable in the wall. Insert your coat hanger through the hole in the wall and swing it back and forth until you catch the cable. *Figure 7-6* shows a wall section and illustrates "fishing" for the cable.

5. Once you grab the cable with the coat hanger, pull it out of the wall. Pull enough cable through the hole to install a male F connector on the end. Connect the F connector to the wall plate outlet. Mount the wall-plate outlet to the outlet box as shown in *Figure 7-5*. A separate cable is used between the wall plate outlet and your TV or cable converter. The outlet box installation was shown in *Figure 4-2*.

Attaching a wall plate outlet to drywall is not very permanent. It can be done by using plastic inserts into the drywall to secure the screws that hold the wall-plate outlet.

6. For the final step, go back into the attic and run the other end of the cable to the splitter. Dress it down, attach cable clamps if necessary, and attach an F connector to the end of the cable. Then connect it to one of the ports on the splitter. It is necessary to do this procedure last. This is so you will have enough slack in the line to reach your room outlet. If you do this step too soon, you may not have a long enough cable to complete the connection downstairs. *Figure 7-4* shows how to install an F-connector.

Attic Installation — With Fire Blocks

Walls with fire blocks are most often found in older homes. A fire block prevents a fire inside the wall from spreading to the upstairs rooms and attic. A fire block is a 2″ x 4″ board mounted half-way down and inside a wall as shown in *Figure 7-7*. When a fire block is present, it is difficult but not impossible to run a cable down the wall.

Additional Tools and Material Needed
1. A flex drill bit 3/8″ diameter and 6′ long
2. A small 1/4″ nail
3. Fish tape
4. String
5. Paddle bit (drill bit) 1/2″ to 1″

1. Once in the attic, locate the wall in which you want to run the cable and the 2″ x 4″ top plate. Using the 1/2″ to 1″ wide paddle bit, drill a hole in the top plate directly above the area where you want the outlet located.
2. As shown in *Figure 7-7,* take your 6′ flex bit, insert it into the hole, lower it down the wall until it hits the fire block in the middle of the wall, and drill a 3/8″ hole down through the fire block. Now you have two holes, one through the top plate and one through the fire block.
3. Insert a stiff fish tape through the top plate hole, push the fish tape down through the hole in the fire block, and down until it hits the bottom of the wall. Push enough fish tape through the hole to reach the opening at the bottom of the wall.
4. Now go down into the room where the wall-plate outlet is to be installed. Make a small hole in the wall where the wall-plate outlet will be. If you are installing an electrical outlet box for holding the wall-plate outlet, cut the hole for the box. If not, make a hole large enough to pull coaxial cable through, but small enough to cover with a wall plate. The hole should be opened at the same height and aligned with other wall plates on that wall (about 12″ above the floor).
5. Locate the fish tape that was pushed down from above and pull it through the wall-plate outlet opening. Install an F connector to the end of the coaxial cable to be connected in the attic. Secure the cable to the fish tape.

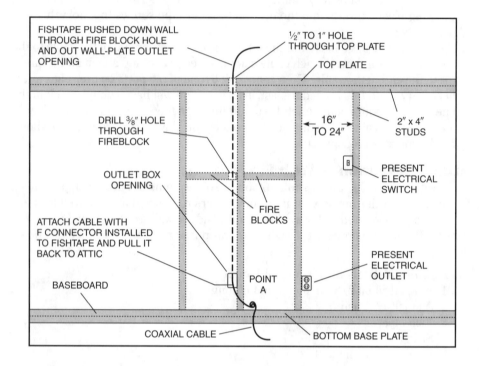

Figure 7-7. Installing Wall-Plate Outlet in Wall with Fireblocks.

6. Go back to the attic and pull the coaxial cable up through the two holes to the attic. This will be the most difficult part of the installation. Be prepared to spend anywhere from 30 minutes to an hour. Have plenty of water with you, for attics can become quite warm even in cool weather.

7. Connect the coax cable to a splitter in the attic that is connected to the cable from the ground block.

8. Go back downstairs, cut the coax cable to the proper length, and install a male F connector. Connect it to the wall-plate outlet, and mount the wall-plate outlet just as in *Figure 7-5.*

PLANS FOR AN ENTERTAINMENT CENTER?

Are you designing an entertainment center for your new house? Do you like the convenience of having the stereo, CD player, VCR, and video system all in one spot? Most people will answer that question with a resounding *yes!* For this reason, we would like to suggest the beneficial arrangement shown in *Figure 7-8.* We suggest grouping a phone jack, an electrical outlet (120 VAC), and a cable outlet. The phone jack is necessary to order pay-per-view programs; the electrical outlet is necessary for power; and the cable outlet (hardline, wireless cable, big-dish satel-

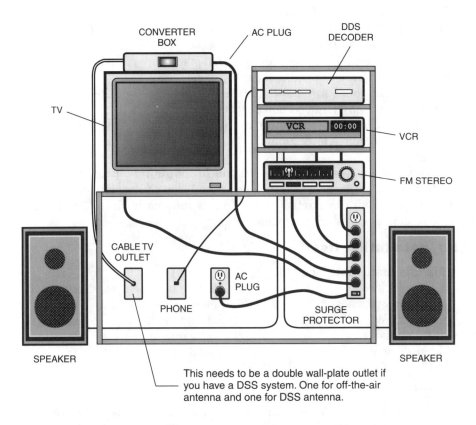

Figure 7-8. Group Cable TV, Telephone and Power Outlets Together for Entertainment Centers.

lite or DSS) is for the TV connection. Ideally, as shown in *Figure 7-8,* they should be positioned side by side and centered behind the entertainment system. This suggestion is particularly important If you are building a new home. Tell your builder to put the outlets together. If you are a do-it-yourselfer, get busy! Grouping these three outlets together will make your hookups tidy and convenient.

A Word of Caution

There are no early warning devices to tell you that a power surge is on its way. It only takes a split second for 6,000 volts to travel from the utility pole to your $2,000.00 entertainment center. Before you connect AC power, install a surge protector between your entertainment center's main power cord and the wall outlet. This will reduce the likelihood of damage from a power surge. Computer, electronic equipment, and electronic parts stores have a wide variety of surge protectors from which to choose. Some cost as little as $7.00. It's a small price to pay to guard such a major investment.

INSTALLING A HOME SATELLITE SYSTEM

The home satellite (big dish) system has the greatest variety of programming of all the TV video systems available. But it will not cover your local TV stations. You still will need an off-the-air TV antenna for local stations. Before you purchase a satellite system from a dealer, consider the following questions:

1. **Do you have room for a satellite dish?** The home satellite dish is a large unit. It can take up a large portion of space in your backyard.
2. **Do you have a place to mount the satellite dish?** Erecting a pole for the satellite dish requires level ground or a place to dig a hole to set the mounting pole in cement.
3. **Do you have access to tools to aid you in mounting a pole?** You will need tools to mix the cement, a level, maybe a plumb line and a compass.
4. **Do you have an obstructed view of the southern horizon?** Tall trees and large buildings can block reception.

The Installation Itself

Material and Tools Needed

1. Screwdriver
2. Socket wrenches
3. Electric or cordless drill
4. Compass
5. Spade or shovel
6. Mixture of cement, sand, and gravel
7. Level
8. Silicone caulk
9. 3/8″ mortar drill
10. Wall-thru tube
11. F connectors

Satellite Dish Placement

Find a good place to set the mounting pole for the dish. Make sure the area is free from large obstacles blocking the view of the southern horizon. It is best to use a compass to verify which way is South. An inexpensive compass, such as one you would use for hiking or driving is adequate to the task. *Do not trust your memory to reference where South is.*

Setting the Mounting Pole

Dig a hole about two feet deep, and about one foot across. Set the pole into the ground in the center of the hole so it will stand up alone. Obtain several bags of cement already mixed with sand and gravel at your local home improvement center. Prepare the cement. Follow the directions for mixing on the bag. Pour the cement into the hole. Using a level, make sure the pole is perfectly perpendicular to the ground, and check it again after the cement is poured. Allow 72 hours (three full days) for the cement to dry.

Dish Assembly and Mounting

Assemble the satellite dish by following the owner's manual. Be careful not to damage the feedhorn. The feedhorn is the device sticking out from the center of the dish. Attach the mounting assembly to the top of the mounting pole.

Once assembled, connect the dish to the mounting plate on top of the mounting pole. Secure all mounting bolts firmly, as this is where the majority of the weight rests. If your satellite dish is motorized, play close attention to the manufacturer's instructions. Be aware that in order for the steering unit to function properly, you may have to have the dish set to a specified location at the start. This will be the point of origination the steering mechanism will use to get its bearings, so read your instructions and use your compass! Missing this initial setup will mean the system will not position the dish correctly. Also, make sure the dish assembly is grounded to a ground stake.

Wiring

After the dish is assembled and mounted, connect a house drop to the feedhorn. The feedhorn collects the information, converts it to a lower frequency, amplifies the signal, and sends the signal through the house drop. A house drop is the cable which brings the signal into your home to the receiver. It is a coaxial cable that has low loss at the down converted frequency.

CAUTION
When drilling through walls make certain there are no electrical wires or water pipes in the way.

Follow these steps to install the house drop:
1. To get the house drop into the house, you will need to drill a hole using a 3/8" mortar drill bit 12" long through the exterior wall. A convenient place maybe right behind where the TV is going to be located.
2. Install a coax feed-through bushing or "wall-thru" tube in the hole that was drilled. Either one has plastic covers made to seal the hole on the exterior wall. They are most effective with silicone seal as a caulk.
3. Dress the cable from the antenna to the house the way it is supposed to lay, then push the end without a connector through the feed-through tube. Seal the tube after feeding all the cable through.
4. On the inside of the house, slide a wall plate over the cable and mount the place to the wall. Make sure plastic inserts are used for the screw mountings.
5. Attach an F connector to the end of the cable and connect it to the input port on the satellite receiver. The satellite receiver is the last major component of the system. It is a specialized type of converter box.
6. The final step is to connect a cable from the output port of the receiver to the TV.

If you want satellite reception on more than one TV, you have to attach a splitter at the *output* of the satellite receiver. From the splitter, you can run lines to any other TVs in the house. If you can to receive local off-the-air TV station, you will have to install an "off-the-air" antenna. See the special section on off-the-air antennas.

WIRELESS CABLE TV INSTALLATION

Wireless cable TV (MMDS) is in a limited number of communities. It requires a tedious and highly technical installation process. Because of this and because parts are difficult to obtain, it is not likely that many consumers will undertake this project on their own. Although illegal MMDS systems are available by mail order, to receive signals with a clandestine dish is piracy. The MMDS antenna must be precisely aimed to receive a sufficient level of signal. If you do not have the correct instrument for receiving a microwave TV signal, you will never be able to aim the dish properly. We highly recommend that you schedule an appointment and let the wireless cable TV company install the system using *their* equipment. Refer to Chapters 1, 3 and 5 for system details.

INSTALLING A DIGITAL SATELLITE SYSTEM

There is one important issue to consider before purchasing an 18″ Digital Satellite System. The area where the dish will be mounted on your house will need an unobstructed view to the South. If you live in a heavily wooded area with tall trees, you may want to reconsider the purchase. The same is true if you live in an area with many tall buildings blocking your view to the South. If you are faced with either of these situations, consult with your DSS dealer. He may have an alternative system that is better suited for your home. In addition, an off-the-air antenna system is necessary to receive all the local channels in your area. Digital Satellite Systems do not receive local UHF or VHF TV signals. You, of course, can use your VHF, UHF system that you now have. See the special section on off-the-air antennas if you need to install one.

Let's Get to Work!

Since you have decided to buy a DSS system and install it yourself, hopefully the following steps should help you do it successfully. If you have traditional hard-line cable in your home you can continue to use it and hook it into your DSS system, or it is possible to use the existing cable TV outlets for distribution of your DSS system signal. If you do use the existing hard-line system, there is the possibility the coaxial cable is of an inferior grade and may have some effect on the quality of the DSS signal. The following steps are based on installing new cable and new wall-plate outlets:

1. Mount the 18″ satellite dish in a secure area on or around your house. The most common places to mount a digital satellite dish are on the roof, on the side of the house, or on the chimney. A ground mount (a level area where it mounts on a pole) may be considered, but the DSS dish is so mall that it would never be mounted directly on the ground like a "bid-dish" satellite system. Connect the coaxial cable to the antenna and run it down the roof, under the overhang to protect it from the weather, and to a ground block. Hold the cable in place with coax nail-in clips or insulated staples.

2. Mount a new ground block on the outside of the house. Run a ground wire to a ground stake. Connect the coaxial cable coming from the digital satellite dish to the ground block. It is very important to make sure there are no splitters between the ground block and the satellite receiver (converter box)! There is an AC current that must travel from the DSS receiver to the satellite dish, without any interruptions. A two-way splitter would block the flow of the AC current.

Decide how the coax cable will be installed from the ground block to the wall-plate outlet in the room where the DSS receiver is to be located. Estimate the length of cable needed and purchase it (RG-6 preferred), and then follow the previous directions for the type installation—crawl space below the house, attic, exterior, etc. Hopefully, the old hard-line cable outlet will be nearby so the cable from it to the DSS receiver will be relatively short. This outlet, as shown in *Figure 7-9*, can be used for the off-the-air TV input if not used for hard-line cable.

3. After the wall-plate outlet has been installed, make the remaining connections as shown in *Figure 7-9*. Connect a coax cable from the DSS wall-plate outlet to the SATELLITE INPUT terminal and a cable from the off-the-air TV input signal wall-plate outlet to the IN FROM ANT terminal on the DSS receiver. For the satellite system and an off-the-air antenna, an A/B switch is required to switch between the two signals. (See *Figure 5-4*.) In the DSS system, the A/B switch is built into the DSS receiver.

4. Connect the phone cable or wire included with your DSS kit from the back of the satellite receiver (converter box) to a phone jack. If you do not hook up the receiver to the telephone line, you cannot receive the pay-per-view channels. When you do want to watch a program on pay-per-view, the instruction and other information is given over the telephone line, and the programmer (DirecTV™ or USSB™) will charge you accordingly. Suppose a phone jack is not available, yet you want one near the TV: PROMPT Publications has a book, *The Phone Book*, which gives you the information you need for phone line installations.

5. There are other connections indicated on *Figure 7-9* to include a VCR and to split the signal to another TV. Also, S-VIDEO and AUDIO R and L connections can be made for superior quality and audio if your TV set has these connections.

6. In most DSS receivers, an access card must be in place that determines system security and access programming. Make sure that the card is plugged in. The system will not work without it.

7. Now you are ready to turn on the DSS receiver. Never plug in the digital satellite receiver until you have totally completed the installation. Some DSS receivers have a programmed-in self-test, which is usually run initially to make sure the system is set up properly. Station selection and program selection is done by menu selection on the screen. Consult the owner's manual for details.

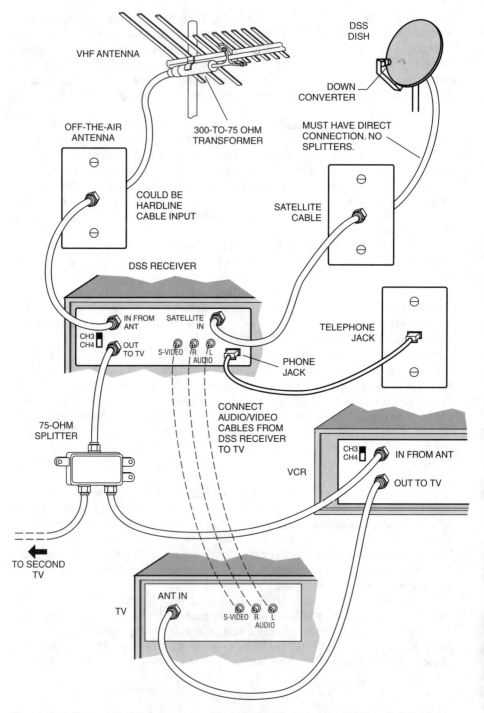

Figure 7-9. Complete DDS System with Off-the-Air Antenna, VCR and Splitter to Second TV.

Off-the-Air Antennas

Home satellite systems and Digital Satellite Systems require an off-the-air antenna to receive local stations. The off-the-air antenna can be mounted as any TV antenna, either on the roof with guy wires, strapped to a chimney, or installed in the attic. A typical antenna is shown in *Figure 7-10*. Place a 300-to-75-ohm transformer at the antenna, attach a 75-ohm coaxial cable to the transformer and run the cable to a ground block. From the ground block, install the cable to bring the antenna signal to a wall-plate outlet. Use the installation steps described previously for the respective types of installation. Should you want an off-the-air antenna, your local electronics distributor can help. A knowledgeable sales assistant can show you the best off-the-air antenna system for your area.

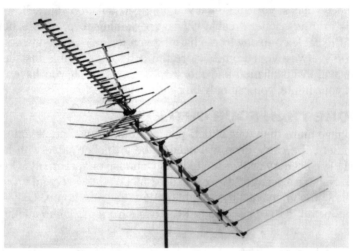

Figure 7-10. A Typical UHF/VHF TV Antenna for Reception of Local TV Stations

WHAT'S LEGAL AND WHAT'S NOT

There are basically two rules when questioning the legality of do-it yourself hardline TV installations. Is it the cable company's property or is it your property. Here are guidelines for you to follow:

1. If it is cable company's property, the general public is not allowed to tamper with the equipment.

   a. A house drop, the connecting link from the cable company's tap to the ground block on your house, is *off limits.* You cannot connect your own house drop to get free cable TV. That's illegal. If the original house drop is damaged by an accidental cable cut or by animals chewing through it, contact your local cable company and let *them* make the repairs.

   b. You cannot tamper with pedestals, feeder lines, amplifiers, and filters that are the cable company's property. The handling of these devices is restricted to the cable company's service technicians

   c. All converter boxes are the cable company's property, and are sealed to discourage subscriber tampering. If you are tempted to open one and modify it to steal premium cable service, you are likely to do more harm than good. You will likely damage the converter box which will result in a costly service charge from your cable company. Theft of service is also *against the law*, and you may be vulnerable to state or federal prosecution.

    d. *Never* use any kind of RF amplifier to overcome low signal problems. RF amplifiers can cause spurious signal leakage that can interfere with aviation, amateur, commercial, and public service communications. This leakage can be detected by the cable company. FCC rules give the cable company the right to shut off service to any house that is causing major leakage.

2. Your installations can begin from the ground block on your house, or from the feedline into your apartment (usually the wall-plate outlet.)

    a. It is now legal to use a splitter in your home, and add up to three additional outlets. This will cost you nothing if you do the installation yourself.

    b. This is a new federal cable TV law pertaining to pre-wired homes that prohibits the cable company from charging you for more than one outlet; therefore, if you pre-wire your home before the contractor's cable installer arrives, you will save on installation and monthly service fees. If you have up to four outlets in your home, you can only be charged for one.

NEED MORE THAN FOUR OUTLETS?

You cannot have more than four outlets per incoming cable line. If you want more than four outlets, the cable drop will need to be a thicker cable with lower loss, called RG-11. The primary cable used to install a house or apartment is RG-6. Or an alternative is to have a second RG-6 cable drop installed to a second ground block. This will allow you to have up to eight outlets in your home. Only the cable company can install RG-11 or a second RG-6 cable from the tap to another ground block.

RENTERS BEWARE

If you are a renter and desire to add additional outlets, here are a few items to consider. First and foremost, if you live in a rented home, get permission from the owner before starting any wiring project. It is possible that the landlord already has a private contractor on call to perform this kind of wiring. If permission is not obtained, it is possible you will lose your security deposit, or incur a damaged property charge. Always ask first! In the case of apartments and duplexes, you may not be able to add additional outlets due to lack of access to the crawl spaces. If you live on the first floor of a multilevel apartment, there is no attic. Even if you do have an attic, the apartment management may consider it taboo to enter the attic! The best advice is to check first before you do anything!

SUMMING UP

We've covered a lot of territory in this chapter. In addition to step-by-step instructions on how to wire up your home for different types of video systems, we've passed along many important tips. Remember to always use the right tools and equipment for the job. Plan ahead for all wiring jobs. Don't be afraid to ask questions before starting.

Chapter 8
Adding Accessories

In today's world, the TV set seldom stands alone in your entertainment area. In this chapter, we discuss some of the accessories you can add to enhance your video system. And we show you lots of different ways to connect them, too.

COMMON SYSTEMS
It doesn't do any good to have the technology if you don't know how to hook it up. The pages that follow contain connection diagrams for cable-ready and non-cable ready systems with TVs, VCRs, and FM-stereo hookups! We even show you how to connect a Closed-Caption (CC) decoder.

A Variety of Connection Diagrams
Almost every home has a VCR these days. This section contains seven different connection diagrams that will be invaluable when it's time to get behind the TV set and start making connections or changes. Each one includes a VCR and other common signal sources and components. Pick the one you like best, or combine two or more to build your own *custom* system.

Figure 8-1 shows simple connections for when you don't need a decoder box — when your TV and VCR are cable-ready and you are not receiving any scrambled channels. With these connections, you can view and record two different channels, or you can monitor the same program you are recording. Use the TV or VCR to tune the channel for viewing (TV on Channel 3 or 4 for VCR tuning), or, when viewing and recording separate channels, use the TV tuner for channel viewing and the VCR tuner for the recording.

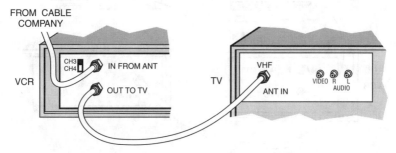

Figure 8-1. Connections for Cable-Ready TV/VCR and Unscrambled Cable Channels.

The basic connections shown in *Figure 8-2* are for a TV, VCR, and cable decoder box when all your cable channels are scrambled. Set the TV and VCR to the decoders's output channel (3 or 4) and use the cable box to change channels. Because the cable box can select only one channel at a time, you cannot watch and record different channels.

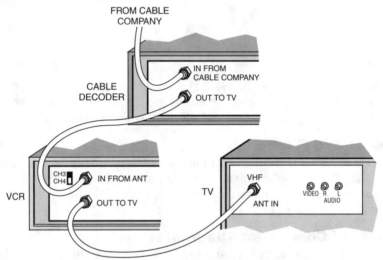

Figure 8-2. Basic Connections for Scrambled Cable Channels

Figure 8-3 shows the addition of a Closed-Caption decoder to the basic scrambled cable system. You'll be able to record the closed captions using your VCR, but you cannot watch and record different channels.

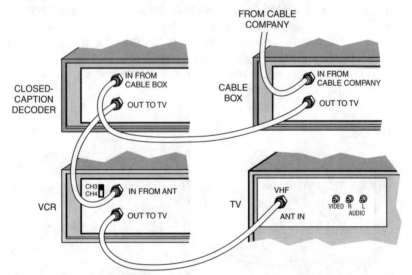

Figure 8-3. Adding a Closed-Caption Decoder to Scrambled Cable is a Great Option for the Hearing Impaired.

If your TV is cable-ready and you receive at least a few unscrambled channels, you'll be able to record any cable channel (scrambled or not) while watching a different unscrambled channel using the connections in *Figure 8-4*. The A/B switch lets you select between the unscrambled cable channels and the output of the VCR for your TV. When you want to watch scrambled channels, you'll have to get them from the VCR's output. Of course, you could add a Closed-Caption decoder to this system, too (between the cable wall outlet and the 2-way splitter).

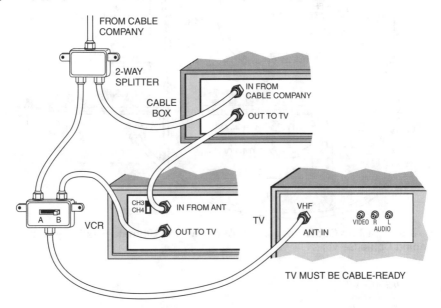

Figure 8-4. An A/B Switch Adds Flexibility to Partially Scrambled Cable Service, If the TV is Cable-Ready.

If most or all of your cable channels are scrambled, the connections in *Figure 8-5* offer the maximum flexibility. You'll need two cable boxes; one to feed the TV and another to feed the VCR. The A/B switch lets you switch between cable channels and video tapes (from the VCR).

Figure 8-6 is for the apartment dweller (or anyone who lives in a strong signal area). These connections let you record a cable channel while viewing a local broadcast channel off-the-air with a pair of 'rabbit ears.' If your cable channels are scrambled, you'll need to add the cable box between the cable outlet and the VCR's input. Use the A/B switch to select between broadcast and cable channels (through the VCR) for the TV.

Note:

■ If broadcast signals are weak, see *Figure 8-9* for the connections for an outdoor antenna instead of rabbit ears. You may not need the booster amplifier.

If some of your cable channels are unscrambled, *Figure 8-7* shows another way to hook up your system if your TV is not cable-ready.

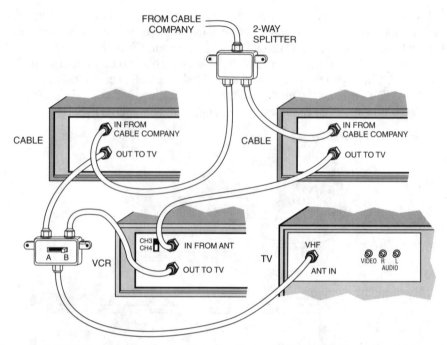

Figure 8-5. With Two Cable Boxes, You Can Have the Ultimate Flexibility with Scrambled Cable Service.

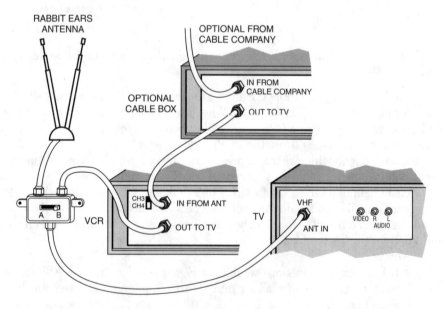

Figure 8-6. "Rabbit Ears" are a Reliable Addition to Cable Reception If You Receive Strong Broadcast TV Signals.

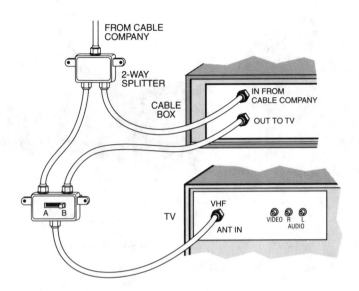

Figure 8-7. With Cable-Ready TV and at Least Some Unscrambled Cable Channels, an A/B Switch is a Simple Way to Select Between Scrambled and Non-Scrambled Programs.

Using an A/B Switch

Throughout the book we often refer to using an A/B switch. (See *Figure 8-8*.) An A/B switch is designed to select between different video sources or to switch one program source between two different video devices (for example, between two TVs, two VCRs, or a TV and a VCR). The most typical systems we refer to that use an A/B switch are the Digital Satellite System and the large-dish satellite system. You can use the A/B switch to select either off-the-air VHF and UHF stations or programs from the satellite system. You can select between off-the-air stations and cable TV, too!

If you have more than two program sources, there are three-way (A/B/C) and four-way (A/B/C/D) switches. You can use an A/B/C switch to select between cable TV, a laser disc player, and a VCR.

As we have mentioned before, always purchase high-isolation A/B switches. These ensure that you do not create signal leakage that can interfere with your own video equipment or even with your neighbor's equipment. If your system is extremely complex, electronics parts stores often carry more sophisticated switches that allow you to independently select from up to five video sources for your TV or VCR. These switches have built-in amplifiers to assure adequate signal strength. And there are even remote control A/B switches, shown in *Figure 8-8b*.

a. A/B High-Isolation Switch   b. Remote Control A/B Switch

Figure 8-8. High-Isolation A/B Switches.

A Word About Closed-Caption Decoders

A closed-caption decoder is a device used to help hearing-impaired people enjoy TV programs. The closed-caption decoder is connected between the signal source and the TV. It translates binary data embedded in the TV signal into visual text that is displayed on the screen along with the picture. As the show progresses, the text appears on the screen line by line. It's called "closed caption" because the text is not visible unless you have a separate decoder or a TV with the closed-caption decoder built in and switched "on." You' might have seen "open" captions if you've ever watched a foreign film that had English subtitles. Even if your hearing is just fine, closed captioning is great if you want to watch TV while a companion wants to sleep.

Even better news, there is some progress in TV design and manufacturing. Many home videos now have closed-caption decoders built in. Look for the Closed-Caption logo — two C's inside a stylized outline of a TV set. *Figure 8-3* shows you how to connect the closed-caption decoder to your TV.

Blacked Out Games? No Problem!

What if your favorite football team blacks out local games and you need an outdoor antenna to receive distant signals? *Figure 8-9* should grab your attention! It is similar to the connections in *Figure 8-6,* but uses an outdoor antenna and booster amplifier instead of indoor rabbit ears. It is critical that you install the booster amplifier *exactly* as shown. If you don't, the booster can create signals that will interfere with your own reception and possibly your neighbors'. Also, be sure to follow any special instructions that come with the amplifier. With a large antenna and a powerful signal amplifier, this installation will be great in rural areas. Plus, you'll have a real performer should your cable service be temporarily interrupted.

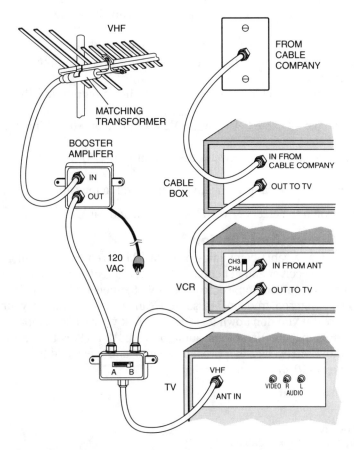

Figure 8-9. A System to Switch Between Hardline Cable and an Outdoor Antenna. The Antenna Booster Amplifier is Added (if Necessary) Used to Amplify Weak Stations.

Actual Installation

Since this is likely to be a very useful system for many readers, we have included a list of materials needed and the installation steps.

Materials and Equipment

Description

Four short coaxial cables (lengths depend on the
    locations of your equipment)
Outdoor Antenna
Booster Amplifier
A 300-to-75 ohm Matching Transformer
(These are sometimes
    supplied with the TV antenna)
A long run of coaxial cable from the antenna to the TV

1. Connect the 300-to-75 ohm transformer to the TV antenna; then connect one end of the long coaxial cable to the transformer and run the cable inside to the amplifier.
2. Connect the other end of the long coaxial cable to the amplifier's input.
3. Connect a short coaxial cable between the amplifier's output and the A input on the A/B switch.
4. Connect a short coaxial cable between the cable box's output and the VCR's input.
5. Connect a short coaxial cable between the VCR's output and the B input on the A/B switch.
6. Connect a short coaxial cable between the output of the A/B switch and your TV.
7. Be sure the wire from the cable company is connected to the cable box's input.

   Use the A/B switch to select between the antenna and cable channels.

Cable to FM-Stereo Installation

Did you know that many cable companies provide FM-stereo programming? You can hook up your FM receiver and get crystal-clear stereo reception without installing an outdoor antenna. In addition, some cable TV channels transmit their stereo audio signals on FM stereo frequencies so you can pick them up on your FM receiver! This is great if you don't own a stereo TV. Ask your cable company for a list of cable channels that *simulcast* their audio on FM. All it takes to connect the cable signal to your FM receiver is a two-way splitter and a few additional feet of coaxial cable. *Figure 8-10* shows you how to do it.

Materials and Equipment
Description
Splitter (two-way)
Two or Three short coaxial cables
   (lengths depend on the locations of your equipment)

75-to-300 Ohm Matching Transformer (if your FM
   receiver has a 300-ohm antenna connection)

1. Connect the cable TV wire that comes from your wall to the input of the two-way splitter. (If you already have cable TV, you'll have to disconnect the cable from the input on the cable box.)
2. Using a short coaxial cable. connect one of the splitter's outputs to the cable box's input.
3. Using another short coaxial cable, connect the splitter's other output to the antenna input on your FM stereo receiver.
   Notes:
   - If there is already an antenna connected to the FM receiver, disconnect it first. Having two signals connected to the same receiver can cause severe interference.
   - If your receiver has a 300-ohm FM antenna terminal (two screws), insert a 75-to-300 ohm matching transformer between the cable and the terminals.

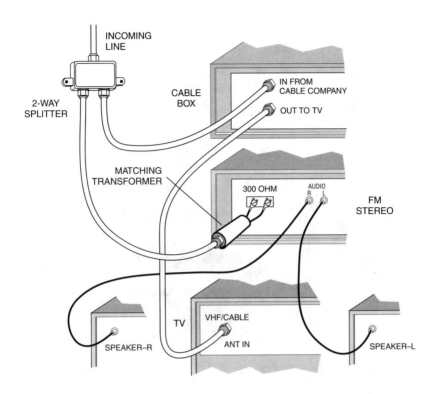

Figure 8-10. Adding FM Stereo Receiver to Cable.

4. Connect a third coaxial cable between the cable box's output and the VHF/ CABLE input on your TV. (If you already have cable TV, this connection already exists.)

Now simply turn on your TV and FM receiver and tune them to the corresponding channel and station.

Picture-In-Picture

Picture-in-Picture (PIP) is a TV feature that has become a great advance in TV viewing. PIP allows you to view two different sources of programming *at the same time on one TV set*! View two different television stations at the same time, or watch a video tape from the VCR while keeping up with the latest NFL game! One program appears in a small window in the corner of the screen while the other program fills up the rest of the screen. And you can switch the two programs between the main and secondary pictures any time you want; you hear only the sound for the main program. This adds a whole new dimension to sports viewing! To get the most out of your VCR and a TV with PIP, connect your system as shown in *Figure 8-11.*

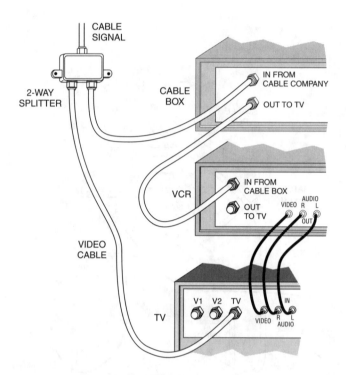

Figure 8-11. Picture-in-Picture Installation.

Materials and Equipment

Description

Splitter (two-way)
Three short coaxial cables (the lengths depend
    on the locations of your equipment)

Audio/Video cable with RCA-type connectors

1. Connect the cable TV wire that comes from your wall to the input of the two-way splitter. (If you already have cable TV, you'll have to disconnect the cable from the input on the cable box.)
2. Using a short coaxial cable, connect the cable box's output to the VHF input on the VCR.
3. Using a second short coaxial cable, connect one of the splitter's outputs to the cable box's input.
4. Using a third short coaxial cable, connect the splitter's other output to the antenna input on your TV.
5. Using shielded video cables with RCA connectors (sometimes called *phono* connectors), connect the VCR's audio and video output jacks to the TV's corresponding audio and video input jacks.

With this system, you can use the VCR's tuner to select a scrambled cable channel for recording and/or viewing and select the second PIP channel using the TV's tuner. If all your cable channels are scrambled, you'll need two cable boxes to get the most out of the system. Or you could connect the cable box between the splitter and TV and use PIP to watch a video tape and a cable channel at the same time.

HOME THEATER SYSTEMS

Home theater systems are becoming more popular with the advent of affordable large-screen TVs. Once you purchase one, you will experience a new dimension in TV viewing. The specification to consider for a home theater TV is at least a 25" screen with a stereo tuner. This type of TV not only provides a spectacular video display, but typically has all the inputs and outputs necessary to add the extra equipment required to make it true "home theater" quality!

A key item to add is a "surround sound" stereo receiver and speakers. The receiver should have inputs for a TV, a Hi-Fi Stereo VCR, a CD player, and a cassette recorder. A surround sound receiver will have all the essential connections, audio output power, and features. The front pair of stereo speakers should be full-range types with sufficient power-handling capability to work efficiently with the receiver. *Figure 8-12* shows a top-of-the line surround sound receiver. An optional subwoofer will deliver the thundering low-frequency sound that adds a tremendous element of realism to popular programs and videos. Indulge yourself and add three more speakers to the system. The receiver separates the audio into five different channels. *It's the next best thing to being there!*

A Receiver with Dolby Pro-Logic Surround Sound.

ON-SCREEN TV GUIDE

An accessory available in parts stores that is very useful is called VideoGuide™. It provides TV program listings for the next seven days on your TV screen through a wireless pager network receiver that sits on top of the TV. With the program listings displayed on the screen, you can select a program using a remote control, click on the program, and the proper channel will be selected so you can view the program. What is particularly useful is the selection of programs for recording on a VCR. Just select the program with the remote control, press the record button and the program will be recorded—even up to seven days in advance, and one time or on a regular schedule. On-screen news summaries, sports summaries, and local weather are also available.

TECHNOLOGY DEMYSTIFIED

You're now set to meet the challenge of just about any video installation. Use the diagrams and step-by-step instructions found in this chapter as your guide and you can't go wrong! If a setup you encounter is different from any of the connections presented in this chapter, it probably uses elements of several of the diagrams put together.

We hope that some of the discussions in this chapter have you thinking about more than just stringing cable to all your TV sets. With the revolution in technology, there is equipment available that will turn an evening in front of the TV into a night at the local movie theater—minus the sticky floor and crying babies! Much of it is at a neighborhood electronic store.

The hookup basics in this chapter have laid the groundwork for what we'll be exploring next—troubleshooting and repair. Read on!

Chapter 9
Troubleshooting and Repairing Systems

THE CABLE COMPANY'S EQUIPMENT IS OFF LIMITS

We have stressed many times in this book that certain parts of a cable TV system should be left alone—specifically, the components that belong to the cable company. All the components from the ground block back toward the tap and beyond are under the control of the cable company. This includes the parts we told you about in Chapter 3, such as the cable drop, line extenders, and feeder lines. Similar restrictions apply to wireless cable and leased DSS installations. However, if you have a large-dish home satellite system or you own your own DSS, you are under no restrictions if you want to make changes or repairs. It's yours!

Cable Drop

The cable drop is the cable connected between the cable company's tap on the utility pole and the ground block on your home. This cable drop is off limits. You cannot connect a cable drop to your home yourself, and you cannot repair the existing cable drop yourself if it's damaged. Although it can be time-consuming and inconvenient, you do have to schedule a time for the cable company to come out and install a new cable drop or fix a damaged one. So, if you suspect that cable company hardware is responsible for a problem, call the cable company.

Taps and Pedestals

Not all taps are on overhead lines; some are underground. If you live in an apartment, the tap for your cable service is probably located in a metal box called a *pedestal*. (See Chapter 3.) Or there may be other metal boxes where feeders lines that are run underground are terminated. This occurs particularly in multiple-building apartment complexes. These devices are off limits to everyone except technicians who work for the cable TV company. If you tamper with these items, not only do you run the risk of personal injury, you might damage them and disrupt your own and your neighbor's cable service!

The Cable Converter/Decoder Box

The cable converter ("descrambler" or "decoder") box is another off-limits device. As mentioned in previous chapters, the converter box is the property of your cable TV company. Never disassemble or tamper with the converter box. In some specialty magazines, converter box computer chips that decode scrambled channels are sold. We greatly discourage both the sale and purchase of such chips.

First, the type of chip you need depends on the cable TV system in your area. There are hundreds of chips to choose from, and it's easy to buy the wrong chip and destroy your cable box. Second, and more important, the converter box is the property of the cable company. You are breaking the law if you as much as open a cable converter box. There are numerous seals outside and inside the box that, if broken, show that the box has been tampered with. See Chapter 3 for examples of several cable converter boxes.

WARNING
If you tamper with any part of the cable company's equipment, you might find more trouble than you imagined possible. Recent legislation makes it a Federal crime to tamper with any part of a cable system that is beyond your side of the ground block. If you decide to save some time, and a few dollars, and hook your home up to cable without the cable company's knowledge, you're a cable "pirate!" Chances are you will get caught when the company performs routine maintenance. If caught, you can be prosecuted and receive a fine and jail time. Because the cable company's revenue comes from the sale of cable service, they get mighty upset when they catch a cable pirate. Don't play games—have your cable company do the installation for you.

WIRELESS CABLE RESTRICTED PARTS

There is no reason for the consumer to ever touch any part of the wireless cable TV system (MMDS). The antenna, mast, feedhorn, diplexer, amplifier/down converter, power injector, and converter box (all shown in Chapter 3) are cable company property. There are no user-serviceable parts in these components, and they should have been set up properly at the time of initial installation. If you suspect your dish has been knocked out of alignment by hail, strong winds, or anything else, contact your wireless cable service provider. Have them check it out instead of attempting the repairs yourself.

Adding Splitters

It is okay if you want to add a signal splitter at the output of the converter box so that you can connect the signal to a second TV set. (*Figure 5-5* illustrates how to do this.)

UNIVERSAL TROUBLE-SHOOTING TIPS

No matter what kind of system you have, these three handy tables will serve as good starting places when you have video system trouble. As we have mentioned many times in this book, cables and connectors are the most common causes of performance problems in a video system.

Table 9-1. F Connectors and Cables*

PROBLEM	CAUSE AND CORRECTION
Snowy Picture	■ Fitting not properly crimped. Crimp it again, or install a new F connector.
	■ Insulation not cleaned off the center conductor. Replace the F connector.
	■ Loose F Connector. Tighten all F connectors.
	■ Water inside cable and/or F connector. Replace cable or dry out and seal the connector.
	■ Too many TVs or other components connected to your cable service. Remove some TVs or have cable company install additional service or low-loss (RG-11) cable.
No Picture	■ Braid touching center conductor. Replace the fitting.
	■ Animal chewed through cable. Repair or replace the cable. See "Repairing Coaxial Cable" later in this chapter.
	■ Broker center conductor in cable (caused by crimped cable). Repair or replace the cable. See "Repairing Coaxial Cable" later in this chapter.
	■ Staple through cable and center conductor. Remove staple and repair or replace the cable.
	■ Bent center conductor in the F connector. Straighten conductor and be sure it is properly inserted in the female connector.

*Assumes coaxial cable for signal paths.

Table 9-2. System with Two TVs*

PROBLEM	CAUSE AND CORRECTION
One TV looks bad, Other TV looks good	■ Loose F connectors at TV. Tighten all F connectors.
	■ Connections to the splitter's inputs and outputs are reversed. Connect them properly.
	■ 75-to-300 ohm matching transformer on TV is bad. Replace it.
Snowy Picture on Both TVs	■ Loose F connector between antenna and splitter. Tighten all F connectors.
	■ 300-to-75 ohm matching transformer at antenna is bad. Replace it.
	■ Defective splitter. Replace it.

*Assumes coaxial cable for signal paths.

Table 9-3. System with Three or More TVs*

PROBLEM	CAUSE AND CORRECTION
One TV looks bad, Other TVs look good	▪ Loose F connectors at TV. Tighten all F connectors. ▪ Connections to the splitter's inputs and outputs are reversed. Connect them properly. ▪ 75-to-300 ohm matching transformer on TV is bad. Replace it.
Snowy Picture on All TVs	▪ Loose F connector between antenna and splitter. Tighten all F connectors. ▪ 300-to-75 ohm matching transformer at antenna is bad. Replace it. ▪ Defective splitter. Replace it.
Grainy Picture on All TVs	▪ Too many outlets? You should connect standard cable service to a maximum of four outlets. Only rarely can standard service supply adequate signal strength to five or more outlets. Have the cable company install a special low-loss cable or a second cable.

*Assumes coaxial cable for signal paths.

INCREASING PICTURE SIGNAL STRENGTH

Too Many Video Outlets

The maximum number of outlets you should have with standard cable service is four. If you have more than four outlets in your house, the signal will become too weak to provide an adequate picture on any of your TVs. To solve this problem, the cable company can install a special low-loss cable (RG-11) between the feeder line and the ground block on your home. Remember, RG-11 cable is larger in diameter than the standard RG-6, so it delivers the signal more efficiently. More signal is available to supply to a greater number of devices than when RG-6 cable is used.

Or the cable company can install a second RG-6 cable drop to a second ground block. Then, you can connect some of your TVs/VCRs to one cable drop and some to the other. This results in a stronger signal level for each device and improves the picture quality to each set. Of course, you will have to modify the video wiring in your home to evenly distribute the two cable signals. Only the cable company can install the low-loss cable drop or a second standard cable drop.

No In-Line Amplifier

As has been stated before, you should not use an amplifier to compensate for problems caused by low signal strength in hardline or wireless cable systems. It's up to the cable company to supply adequate signal strength. Amplifiers can cause signal leakage that creates interference to your own and your neighbor's reception. Leakage can also interfere with vital public service and aviation communications.

Federal Communications Commission (FCC) rules give the cable company the right to shut off service to any customer that is causing major interference.

PREVENTING INGRESS & EGRESS

Ingress is a fancy term describing the penetration of external signals into your cable system, resulting in interference to the video and audio. Signal leakage from your system that causes interference for someone else also has a fancy name—egress. It is important to note that any condition that permits other signals to leak into your system can also permit signals from your system to leak out and interfere with someone else's system. The best way to prevent such interference or leakage is to be sure coaxial cables are used and all connectors are properly installed on the cables and are securely connected to their terminals. Also, be sure that the cable's shielding is not cracked or broken anywhere along its entire length.

300-Ohm Twin Lead from Antenna to TV

There are still TV video systems where the signal from the TV antenna is coupled to a TV set with 300-ohm twin lead. Many installations were made this way because twin-lead was inexpensive, had relatively low loss, was easy to work with and didn't require any special connectors. If you lived in an urban or suburban area, most stations were strong enough to be received with acceptable pictures. The one major shortcoming of twin lead—the lack of shielding—didn't bother people until recently. Along with the desired TV and FM signals received by the antenna, twin lead picks up undesired signals all along its length from the antenna to the TV set. That's why you will see twin-lead twisted as it runs from antenna to the TV. The twisting reduces noise pickup because the noise signal is induced in both wires of the twin-lead and the twisting causes cancellation. In addition, the lack of shielding works both ways. Signal carried by twin lead will radiate from it. If signals are quite strong, they will radiate along the length of the twin lead. For this reason coaxial cable has become so essential for today's sophisticated video setups. If installed properly, coaxial cable will not receive any interference (nor will it leak any interfering signals). The bottom line is, if there is any twin lead in your home, do not use any of it when you install your new video hardware. If you do, you will likely not have the picture performance you desire.

Improperly Installed Connectors

Ingress occurs in three main areas—improperly installed F connectors, loose connections, and exposed or chewed cable. The possible leakage effect on a TV picture of an improperly installed "bushy" F connector or when cable shielding is broken and the center conductor is exposed, is shown in *Figure 9-1*. A "bushy" fitting (as shown in *Figure 6-5*) allows signals to leak in and out. Such a fitting most often happens when someone uses pliers to crimp an F connector instead of the proper crimping tool. In *Figure 9-2*, we compare a properly installed F connector to a bushy one. *If leakage from your system causes serious interference in someone else's system, the cable company can discontinue your service until the problem is fixed.*

Figure 9-1. Snowy Picture Caused by Poorly Installed "Bushy" F Connector.

Bushy Fittings

To help you understand why it is important to correctly install connectors, consider the following scenario. A cable run goes through the attic, down the inside of a wall and to your living room TV. Some time passes and you decide to add a splitter so you can enjoy cable TV in your bedroom. You go into the attic, cut the cable and add F connectors. But for some reason you are in a hurry and you crimp the F fittings with a pair of pliers instead of a crimping tool. This results in bushy fittings like the one shown in *Figure 9-2*, and snowy and noisy pictures on both TV sets. However, let's assume that you got lucky and the pictures aren't all that bad. Later that weekend you are set to watch the "big game" on TV. You invite some of your friends over, order pizza, and the next thing you know, you have a real football-watching party going. Then *it* happens! Your lonely, telephone-addicted teenager latches on to the cordless phone for a few hours of quality time with that special someone. The RF radio waves from the telephone penetrate the cable due to the bushy connectors. Instead of gridiron warfare on the TV, you have garbled sound, ghost images, and snow. The party is disrupted, and the great picture you enjoyed for so long is but a memory. The moral to this story is, every time you cut the cable to add connectors, make sure the connectors are installed properly. If the job is done poorly you take a chance of poor picture quality.

When the braided shielding is not properly inserted into the F connector, it is possible for a strand of the braid to touch the center conductor. If this happens, there is no picture at all.

Radio Interference

Another potential problem associated with bushy fittings, or an exposed center conductor due to abrasion or animal chewing, is interference from nearby 2-way radios (such as Citizens Band, public service and amateur radio) or from a nearby

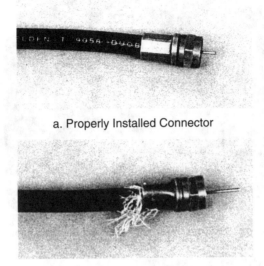

a. Properly Installed Connector

b. Improperly Installed "Bushy" Connector

Figure 9-2. Improperly Installed Connectors Produce Poor TV Pictures.

FM radio station. The frayed braid allows strong nearby signals into your cable installation, and this may create problems with your reception. Most often, this interference is created by front-end overload to your television set, not by spurious signals from the nearby transmitter. In case you are plagued by interference and you are certain your cable installation is flawless, you can buy filters that are easily installed and will possibly eliminate the problem. CB/Ham interference filters often can be installed at the back of each set. FM interference filters filter out that "herringbone" pattern created by FM interference. Always keep in mind that if you have cable television, your service provider will help you with interference problems, no matter what the source.

We point out again, when the braided shielding is not properly inserted into the F connector, it is possible for a strand of the braid to touch the center conductor. If this happens, there is no picture at all.

Effects of Loose Connections

Figure 9-3 shows the effect of loose connections on the TV picture. Loose connections can occur at a splitter, ground block, decoder box, or any other device in the system, including the TV set. Fortunately, this is a problem with an easy solution. Simply check all your F connectors to be sure they are screwed on (or pushed on) as far as they will go. And any time your move around cables or components, check the connections again to be sure these disturbances didn't loosen the connections. A staggering number of video system "problems" are cured by simply tightening a connection or two!

Figure 9-3. Ingress Interference Caused by a Loose F Connector.

Effect of Breaks in Shielding

In Chapter 6, we discussed the problem of animals chewing or gnawing on cable, resulting in exposure of, and/or break in, the cable's shielding. If this occurs, the center conductor is not properly protected and is exposed to ingress of "noise" from electrical appliances or other video signals. And if the center wire is broken, you won't get any signal at all to your TV or the rest of your components. It's a good practice to make sure your coaxial cable does not have the chance to rub or scrape against sharp edges in the attic or crawl space; this can cause the same problems as chewing animals. To prevent this, secure the cable with coaxial cable clips, shown in *Figures 4-6* and *6-4,* or slip a short section of automotive radiator hose over the coax at appropriate locations to prevent the cable from getting damaged. The effects of exposed or broken shielding are shown in *Figure 9-4.*

Figure 9-4. Electrical Interference Caused by a Blow Dryer, Microwave Oven, or Other Appliance When Cable Shielding is Exposed or Broken.

REPAIRING COAXIAL CABLE

In *Figure 7-4* of Chapter 7, we showed how to properly install an F connector. You may need to repair a coaxial cable that has been chewed on or damaged in some other way. If the damage is on the cable company's side of the ground block, you must call the local cable TV company to come out and fix it. If the problem is on your side of the ground block, or if you have your own satellite system or a DSS, you can make the repair yourself. Doing it yourself is faster, cheaper, and easier than hiring someone else to do it, and you get the satisfaction of doing it yourself. Simply refer to *Figure 9-5* and follow these steps to ensure a proper repair.

1. Locate the damaged portion of the cable.
2. Using wire cutters, cut out the damaged portion of the cable.
3. Prepare the freshly cut ends of the remaining cable and attach an F connector to each as outlined in *Figure 7-4*. Be sure there is no braid sticking out of the back of the fittings.
4. The final step is to join the two new F connectors using a F-81 "barrel" connector. Securely tighten each F connector and, if the connections are exposed to the *elements*, seal them with sealant tape.

   Note:

   ■ If you cut out a long section of cable and can't directly join the two new ends, make a patch cable with F connectors on each end and use two barrel connectors to join the sections.

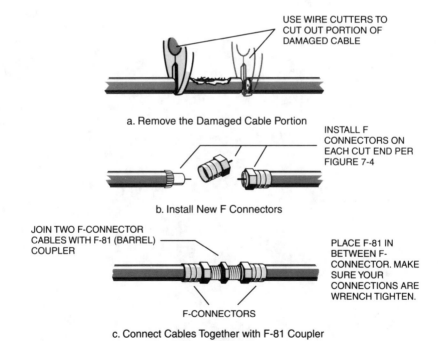

Figure 9-5. Repairing a Damaged Coaxial Cable by Cutting Out Damaged Portion, Installing New F Connectors, and Coupling Together with an F-81 Coupler.

Summary

In this book, we have explained the variety of TV video systems available, how they are installed and interconnected, and how problems can be identified, and repaired. When we began, we were striving to provide clear, concise and easily understood instructions and discussion supported by reinforcing illustrations. That was our goal; we hope we have succeeded.

A Consumer Guide to Product Safety

 This symbol is intended to alert the user of the presense of uninsulated "dangerous voltage" within the product's enclosure, that may be of sufficient magnitude to constitute a risk of electric shock to persons.

 This symbol is intended to alert the user of the presence of important operating and maintenance (servicing) instructions in the literature accompanying the appliance.

CAUTION

RISK OF ELECTRIC SHOCK
DO NOT OPEN

CAUTION

TO REDUCE THE RISK OF ELECTRIC SHOCK,
DO NOT REMOVE COVER (OR BACK).
NO USER-SERVICEABLE PARTS INSIDE.
REFER SERVICING
TO QUALIFIED PERSONNEL.

 Video Products
Important Safeguards

1. **Read Instructions** — All the safety and operating instructions should be read before the appliance is operated.

2. **Retain Instructions** — The safety and operating instructions should be retained for future reference.

3. **Heed Warnings** — All warnings on the appliance and in the operating instructions should be adhered to.

4. **Follow Instructions** — All operating and use instructions should be followed.

5. **Cleaning** — Unplug this video product from the wall outlet before cleaning. Do not use liquid cleaners or aerosol cleaners. Use a damp cloth for cleaning.

6. **Attachments** — Do not use attachments not recommended by the video product manufacturer as they may cause hazards.

7. **Water and Moisture** — Do not use this video product near water — for example, near a bath tub, wash bowl, kitchen sink, or laundry tub, in a wet basement, or near a swimming pool, and the like.

8. **Accessories** — Do not place this video product on an unstable cart, stand, tripod, bracket, or table. The video product may fall, causing serious injury to a child or adult, and serious damage to the appliance. Use only with a cart, stand, tripod, bracket, or table recommended by the manufacturer, or sold with the video product. Any mounting of the appliance should follow the manufacturer's instructions, and should use a mounting accessory recommended by the manufacturer.

9. **Ventilation** — Slots and openings in the cabinet are provided for ventilation and to ensure reliable operation of the video product and to protect it from overheating, and these openings must not be blocked by placing the video product on a bed, sofa, rug, or other similar surface. This video product should never be placed near or over a radiator or heat register. This video product should not be placed in a built-in installation such as a bookcase or rack unless proper ventilation is provided or the manufacturer's instructions have been adhered to.

10. **Power Sources** — This video product should be operated only from the type of power source indicated on the marking label. If you are not sure of the type of power supply to your home, consult your appliance dealer or local power company. For video products intended to operate from battery power, or other sources, refer to the operating instructions.

11. **Grounding or Polarization** — This video product is equipped with a polarized alternating-current line plug (a plug having one blade wider than the other). This plug will fit into the power outlet only one way. This is a safety feature. If you are unable to insert the plug fully into the outlet, try reversing the plug. If the plug should still fail to fit, contact your electrician to replace your obsolete outlet. Do not defeat the safety purpose of the polarized plug.

12. **Power-Cord Protection** — Power-supply cords should be routed so that they are not likely to be walked on or pinched by items placed upon or against them, paying particular attention to cords at plugs, convenience receptacles, and the point where they exit from the appliance.

13. **Lightning** — For added protection for a video product's receiver during a lighting storm, or when it is left unattended and unused for long periods of time, unplug it from the wall outlet and disconnect the antenna or cable system. This will prevent damage to the video product due to lightning and power-line surges.

14. **Overloading** — Do not overload wall outlets and extension cords as this can result in a risk of fire or electric shock.

15. **Object and Liquid Entry** — Never push objects of any kind into this video product through openings as they may touch dangerous voltage points or short-out parts that could result in a fire or electric shock. Never spill liquid of any kind on the video product.

16. **Servicing** — Do not attempt to service this video product yourself as opening or removing covers may expose you to dangerous voltages or other hazards. Refer all servicing to qualified service personnel.

17. **Damage Requiring Service** — Unplug this video product from the wall outlet and refer servicing to qualified service personnel under the following conditions:

a. When the power-supply cord or plug is damaged.

b. If liquid has been spilled, or objects have fallen into the video product.

c. If the video product has been exposed to rain or water.

d. If the video product does not operate normally by following the operating instructions. Adjust only those controls that are covered by the operating instructions, as an improper adjustment or other controls may result in damage and will often require extensive work by a qualified technician to restore the video product to its normal operation.

e. If the video product has been dropped or the cabinet has been damaged.

f. When the video product exhibits a distinct change in performance — this indicates a need for service.

18. **Replacement Parts** — When replacement parts are required, be sure the service technician has used replacement parts specified by the manufacturer or have the same characteristics as the original part. Unautho-

rized substitutions may result in fire, electric shock or other hazards.

19. **Safety Check** — Upon completion of any service or repairs to this video product, ask the service technician to perform safety checks to determine that the video product is in proper operating condition.

20. **Power Lines** — An outside antenna system should not be located in the vicinity of overhead power lines or other electric light or power circuits, or where it can fall into such power lines or circuits. When installing an outside antenna system, *extreme care should be taken to keep from touching such power lines or circuits as contact with them might be fatal.*

21. **Outdoor Antenna Grounding** — If an outside antenna or cable system is connected to the video product, be sure the antenna or cable system is grounded so as to provide some protection against voltage surges and build-up static charges. Section 810-21 or the National Electrical Code, ANSI/NFPA No. 70 — 1984, provides information with respect to proper grounding of the mast and supporting structure, grounding of the lead-in wire to an antenna discharge unit, size of grounding conductors, location of antenna discharge unit, connection to grounding electrodes, and requirements for the grounding electrode. See Figure 1.

FIGURE 1 Example of antenna grounding as per national Electrical Code Instructions

[a] Use No. 10 AWG (5.3 mm[2]) cooper, No. 8 AWG (8.4 mm[2]) aluminum, No. 17 AWG (1.0 m[2]) copper-clad steel or bronze wire, or larger, as ground wire.

[b] Secure antenna lead-in wire and ground wires to house with stand-off insulators spaced from 4 ft (1.22 m) to 6 ft (1.83 m) apart.

[c] Mount antenna discharge unit as close as possible to where lead-in enters house.

[d] Use jumper wire not smaller than No. 6 AWG (13.3 mm[2]) copper or the equivalent, when a separate antenna-grounding electrode is used. See NEC Section 810-21(j).

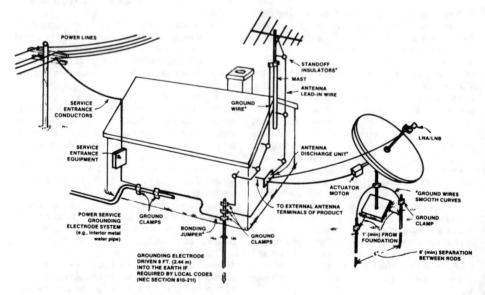

Frequencies of Television Channels

VHF Band			
Low Band		**High Band**	
Channel	**Freq. (MHz)**	**Channel**	**Freq. (MHz)**
2	54-60	7	176-180
3	60-66	8	180-186
4	66-72	9	186-192
5	76-82	10	192-198
6	82-88	11	198-204
		12	204-210
		13	210-216

UHF Band					
Channel	**Freq. (MHz)**	**Channel**	**Freq. (MHz)**	**Channel**	**Freq. (MHz)**
14	470-476	38	614-620	62	758-764
15	476-482	39	620-626	63	764-770
16	482-488	40	626-632	64	770-776
17	488-494	41	632-638	65	776-782
18	494-500	42	638-644	66	782-788
19	500-506	43	644-650	67	788-794
20	506-512	44	650-656	68	794-800
21	512-518	45	656-662	69	800-806
22	518-524	46	662-668	70	806-812
23	524-530	47	668-674	71	812-818
24	530-536	48	674-680	72	818-824
25	536-542	49	680-686	73	824-830
26	542-548	50	686-692	74	830-836
27	548-554	51	692-698	75	836-842
28	554-560	52	698-704	76	842-848
29	560-566	53	704-710	77	848-854
30	566-572	54	710-716	78	854-860
31	572-578	55	716-722	79	860-866
32	578-584	56	722-728	80	866-872
33	584-590	57	728-734	81	872-878
34	590-596	58	734-740	82	878-884
35	596-602	59	740-746	83	884-890
36	602-608	60	746-752		
37	608-614	61	752-758		

A. Logarithms

EXPONENTS

A logarithm (log) is the exponent (or power) to which a given number, called the base, must be raised to equal the quantity. For example:

Since $10^2 = 100$, then the log of 100 to the base 10 is equal to 2, or $\text{Log}_{10}\, 100 = 2$

Since $10^3 = 1000$, then the log of 1000 to the base 10 is equal to 3, or $\text{Log}_{10}\, 1000 = 3$

BASES

There are three popular bases in use—10, 2 and ϵ. Logarithms to the base 10 are called common logarithms (log). Logarithms in base ϵ are called natural logarithms (ln).

Logarithms to the base 2 are used extensively in digital electronics.

Logarithms to the base ϵ (approximately 2.71828...) are quite frequently used in mathematics, science and technology. Here are examples:

Base 10

$\log_{10} 2 = 0.301$ is $10^{0.301} = 2$

$\log_{10} 200 = 2.301$ is $10^{2.301} = 200$

Base 2

$\log_2 8 = 3$ is $2^3 = 8$

$\log_2 256 = 8$ is $2^8 = 256$

Base ϵ

$\text{in}_\epsilon\, 2.71828 = 1$ is $\epsilon^1 = 2.71828$

$\text{in}_\epsilon\, 7.38905 = 2$ is $\epsilon^2 = 7.38905$

RULES OF EXPONENTS

Since a logarithm is an exponent, the rules of exponents apply to logarithms:

$\log (M \times N) = (\log M) + (\log N)$

$\log (M/N) = (\log M) - (\log N)$

$\log M^N = N \log M$

B. Decibels

The bel is a logarithmic unit used to indicate a ratio of two power levels (sound, noise or signal voltage, microwaves). It is named in honor of Alexander Graham Bell (1847-1922) whose research accomplishments in sound were monumental. A 1 bel change in strength represents a change of ten times the power ratio. In normal practice, the bel is a rather large unit, so the decibel (dB), which is 1/10 of a bel, is commonly used.

$$\text{Number of dB} = 10 \log P2/P1$$

A 1 dB increase is an increase of 1.258 times the power ratio, or 1 db = 10 log 1.258.

A 10 dB increase is an increase of 10 times the power ratio, or 10 db = 10 log 10.

Other examples are:

3 dB = 2 times the power ratio

20 dB = 100 times the power ratio

−30 dB = 0.001 times the power ratio

It is essential to remember that the decibel is *not* an absolute quantity. It merely represents a change in power level relative to the level at some different time or place. It is meaningless to say that a given amplifier has an output of so many dB unless that output is referred to a specific power level. If we know the value of the input power, then the *ratio* of the output power to the specific input power (called power gain) may be expressed in dB.

If a standard reference level is used, then *absolute power* may be expressed in dB *relative* to that standard reference. The commonly used reference level is one milliwatt. Power referenced to this level is expressed in dBm. Here are power ratios and dBm ratios:

dB	Power Ratio	dBm	Power (mw)
1	1.258	1	1.258
3	2	3	2
10	10	10	10
20	100	20	100
−30	0.001	−30	0.001

Glossary

A/B Switch: A switch that allows you to change between two or more sources of signal input. (e.g., switch between CATV and regular off-the-air TV signals).

AC Adapter: A power supply that plugs into a regular 120 VAC outlet in the home and is connected to an appliance to supply ac or dc power.

Accessory: A unit attached to or interconnected to a main system to supplement or add to the system's performance.

AC Power Supply: Any source of ac power, such as an ac outlet, transformer, inverter, or ac generator.

Alternating Current (AC): An electrical current that periodically changes in magnitude and in the direction of the current.

Ampere: The unit of measurement for electrical current in coulombs (6.25 x 10^{18} electrons) per second. One ampere flows in a circuit that has one ohm resistance when one volt is applied to the circuit.

Amplifier: An electronic circuit, many times contained in an integrated circuit, used to increase the voltage, current or power of an applied signal.

Antenna: A wire or other conductive metallic structure used for radiating or receiving electromagnetic signals, such as those for radio or video systems.

Array: In an antenna, a group of elements arranged to provide the desired directional characteristics. These elements may be driven elements, reflectors, or directors.

Attenuation: Any reduction in signal strength, usually expressed in decibels.

Band: A frequency range in which radio and TV signals are transmitted over air waves or CATV.

Bandwidth: A specific range of frequencies, from f_1 to f_2, over which the output response

of a tuned circuit, an amplifier, or a total system remains above a specified value.

Barrel Connector (F-81): A threaded cylinder shaped device used to connect two Fconnectors together.

Baseband: Common term used to describe the separate audio and video signals used by VCRs, TV monitors, and many other modern video components.

Beam Antenna: An antenna to which the radiated signal is concentrated in a narrow beam pattern.

Booster Amplifier: A circuit or device used to increase the output current or the voltage capabilities of a signal amplifier circuit without loss of accuracy (ideally) or inversion of polarity.

C Band: The frequency range between 3.7 - 4.2 GHz. The most common use of this band is the downlink frequency for the home satellite system.

Cable (As applies to interconnection): One or more separately insulated wires either bound together or enclosed in a common protective covering. See coaxial cable.

Cable Decoder: commonly called a "cable box," this electronic device converts selected cable TV channels to a channel any standard TV can receive (usually VHF channel 3 or 4). Only cable decoders provided by cable companies are able to decode scrambled cable channels. See Converter.

Cable Ready TV: A TV or VCR that is capable of receiving unscrambled CATV signals without the use of a converter box.

Capacitance (C): The capability to store charge in an electrostatic field. It can be expressed in farads as the stored charge Q, in coulombs, divided by the voltage E, in volts, that supplied the charge. Capacitance tends to oppose any change in voltage.

Capacitor: A device made up of two metallic plates separated by a dielectric (insulating material). Used to store electrical energy in the electrostatic field between the plates.

CATV: An abbreviation for Community Antenna Television, which is called cable TV.

Central Antenna Distribution System: A centralized antenna system which distributes an off-the-air signal to all the apartments in an apartment complex. Also called MMDS.

Channel: A specified range of frequencies that defines the bandwidth assigned for a specific communications link for radio, TV, and other communications to isolate transmissions from each other.

Circuit: 1. A complete path that allows electrical current from one terminal of a voltage source to the other terminal. 2. An interconnection of electrical or electronic components to accomplish a specific function.

Circuit Diagram: A symbolic description showing the component and wiring interconnections of electrical and electronic equipment.

Coaxial Cable: A 4-layer cable with wide bandwidth characteristics used to supply subscribers with a cable signal. This cable consists of copper wire core surrounded by a layer of polyethylene, then wrapped by a thin aluminum or copper braiding, which is covered with a protective vinyl coating. The common types are: RG-59, RG-6, and RG-11. RG-6 is the most common. RG-11 is a larger diameter and lower loss cable. Each is available in single, double, triple, or quadruple shielding.

Compact Disk (CD OR CD ROM): A digital storage medium for music, all types of computer data, and video signals.

Component: The individual parts that make up a circuit, a function, a subsystem, or a total piece of equipment.

Conduit: A protective covering for cable TV lines. Usually it is made of plastic or metal and it prevents the deterioration and weathering of the normally exposed cable lines.

Converter: A device used to convert scrambled cable signals into clear pictures for your TV. This is sometimes referred to as a converter box.

Current: The flow of electrons measured in amperes. There is one ampere of current when one volt is impressed on a circuit with a resistance of one ohm.

Cycle: The pattern of a waveform that repeats itself when a repetitive periodic waveform is plotted on a time axis.

Debug: To detect, locate, and correct circuit problems in a video system.

Decibles (dB): Abbreviation for a measure of one tenth of a Bell. Commonly called a decibel. See Appendix for detailed discussion.

Decoder/ Descrambler: The circuitry inside a converter box that descrambles the cable signal to recover the TV picture.

Device: A hardware or combination of elements that is, or makes up, a component or subsystem. Used commonly to describe a semiconductor transistor, diode, integrated circuit, etc.

Diplexer: A device that combines two or more signals into one signal that is sent down a house drop to a converter box.

Digital: A type of signal consisting of only two signal levels. The levels are generally called on/off, high/low, or 1/0. Because of their simplicity, this type of signal is easy to process and store. TVs and VCRs that have digital converters and processors are able to perform many special effects such as picture-in-picture (PIP). The opposite of digital is analog, which contains infinite signal levels between the highs and lows.

Digital Circuit: Electronic circuits that handle, manipulate, operate on, store, and process information in digital form.

Digital Satellite System (DSS): The latest TV video system that provides the owner / subscriber with up to 150 channels, via a small 18-inch receiving dish.

Drop Cable: The main coaxial cable that couples the cable signal from the cable company's feeder line to the subscriber.

Downlink: This refers to the signal that is being transmitted down from a communications satellite circulating above the earth to a receiving antenna on earth.

Earth Station: These are the satellite dishes that the cable companies use to receive or transmit signals via a satellite.

Egress: To leak a signal into the airwaves from a cable due to a shield break in the cable, poor F-connector installation, or a loose F connector.

Electromotive Force (E): The force which causes an electrical current in a circuit when there is a difference in potential. Synonym for voltage.

Error: Any deviation of a computed, measured, or observed value from the correct value.

F Connector: The standard connector used with coaxial cable for video systems. The VHF and cable inputs and outputs on most TVs and VCRs also use F connectors. F connectors are available in "screw-on" and "push-on" types. The screw-on type are best for outdoor use and for long, heavy cables. The push-on type are best for short cables used for connections between components within your video system.

F-81: See Barrel Connector

FCC: The abbreviation for the Federal Communications Commission.

Feeder Lines: These are the rigid metal lines that distribute the cable company's programming throughout its network

Feedhorn: On an arm in the center of a receiving video system dish antenna, the part that collects the signals that are focused on it by the dish antenna's reflector element.

Fireblocks: This refers to a wall that has a 2″ x 4″ board placed horizontally down the middle of the wall.

Fishtape: A long metal wire with a hook on the end. It is used to pull a cable down and out of the interior of a wall.

Footprint: The area on the earth's surface covered by a satellite signal.

Frame: One complete video picture. There are thirty video frames per second. A frame actually consists of two half-frames called "fields."

Gain: The technical term for amplification of a signal to increase its current, voltage or power level; usually measured in decibels.

Geostationary Orbit: This refers to the average distance a communications satellite is from the surface of the Earth. Satellites are positioned in orbit at a distance of 22,300 miles above the equator.

Ghost: A faint image on a TV screen offset from the main image to the left or right. Usually caused by ingress and other interferences.

Gigahertz: The measure of a frequency's signal, in units of one billion cycles per seconds (GHz).

Ground Block: A device that is used for grounding all the cable wiring in your house.

Hardline: This is the traditional cable TV system distributed to residential and business subscribers. This TV video system is referred to as hardline cable TV due to its delivery system of feederlines and cables from a central cable company location.

HEX Crimper: The hand tool used to fasten F connectors to coaxial cable.

High Band: TV channels that range from 7 to 13. This frequency ranges between 174 - 216 MHz.

Home Satellite System (HSS): The TV video system that provides its owner with up to 250 channels. This system receives signals via its large receiving dish. The receiving dish ranges in size from four feet to 12 feet in diameter.

House Drop: This refers to the coaxial cable, which runs from a tap, or an antenna, to the groundblock or converter box. It is the connecting cable that brings programming into the home.

Impedance (Z): The opposition (measured in ohms) of circuit elements to current in an alternating current circuit. The impedance includes both resistance and reactance.

Inductance (L): The capability of a coil to store energy in a magnetic field that surrounds it. The stored energy tends to oppose any change in the existing current in the coil.

Ingress: Outside sources of interference that penetrate into a video system to distort the video system signal.

Interference: Any electrical disruption of the TV signal which causes distorted images on the TV screen. Interference can be caused from lightning or from machines.

Kilohertz (kHz): A measurement of signal frequency equal to one thousand cycles per second.

Ku-Band: The full frequency range of Ku-band is between 10.95 - 12.75 GHz. This band range is shared by both HSS and DSS systems. The frequency range of 11.7 - 12.2 GHz is received by the HSS antenna. The frequency range of 12.2 - 12.7 GHz is received by the DSS antenna.

Leakage: Usually found at the cable tap, behind the TV, or at damaged cable lines, releasing cable signals into the airwaves.

Line Extenders (LEs): Devices that are used to amplify cable TV signals throughout the cable company's distribution system.

Low Band: TV channels that range from 2 to 6. This frequency ranges between 54 - 88 MHz.

Low-Noise Block Converter Feedhorn (LNBF): A unit on the home satellite antenna arm that is composed of three devices, the feedhorn, the low-noise amplifier and the blockconverter. The LNBF collects the signals and amplifies them 100,000 times, and converted to a lower frequency range that is sent to the satellite receiver.

Matching Transformer: Enables any TV that is not cable ready (doesn't have a co-axial connector for the antenna input, but has screw terminals) to receive a signal from a 75-ohm coaxial cable by converting it to a 300-ohm input for the screw terminals. One side of the transformer is connected to the antenna terminals on the TV and the other side is connected to the 75-ohm coaxial cable. NOTE, this device, in some cases, may protect your TV from a electrical spikes.

Multiple Dwelling Unit (MDU): Living units that contain multiple families, such as apartment complexes and duplexes.

MDS: This is an abbreviation for Multi-point Distribution Service. This refers to the two microwave channels (MDS channel 1 and 2) located in the 2150 - 2162 MHz frequency.

Megahertz: The measure of a frequency's signal, in units of one million cycles per seconds (MHz).

MMDS: This is the abbreviation for Multi-channel Multi-point Distribution Service. The frequency range is between 2500 and 2686 MHz. This frequency contains 28 microwave channels.

Midband: This is the cable TV band range, channels A to I. These channels fall between the off air TV channels of 6 and 7.

Negative Trap: A device that is inserted between the multi-tap and the subscribers TV. It is usually located at the multi-tap. It is used to block out any pay channels. The trap has no effect on the subscribers regular cable reception.

Noise: An unwanted signal or portion of a signal. Audio noise is usually heard as hiss; video noise is usually seen as white sparkles.

NTSC (National Television Standards Committee): A group of businesses and engineers originally created NTSC to decide on early standards for color and black and white television in the U.S. The NTSC system is also used in Japan. Other television standards around the world include PAL (used in most of Europe) and SECAM (used in France, parts of Africa, and the Soviet Union).

Off-The-Air Antenna: This refers to any UHF or VHF antenna that is attached to a TV, for the purpose of receiving local stations 2 - 13 and 14 - 83.

Open Circuit: An electrical circuit in which the current path is incomplete.

Outlet: A wallplate that provides easy access to a connection for cable signal.

Pad: A device used to reduce or attenuate a signal's level by a specified amount.

Parabolic Reflector: Also called a reflector element. One or more conductors or conducting surfaces for reflecting radiant energy from a source into a directed beam. When used for a receiving antenna, it focuses the received energy to the central feedhorn location.

Postive Trap: A device that is placed on the multi-tap or behind the TV. This trap allows the subscriber to receive pay TV channels.

Radiation: The propagation of energy through a medium. In particular, the transmitting of RF energy from an antenna.

RF (Radio Frequency): A high-frequency signal that can be transmitted through the air. such signals are used to carry baseband (audio and video) signals. In television, VHF, UHF, and cable TV signals are considered RF.

Reactance (X): The impedance that a pure inductance or a pure capacitance provides to current in an ac circuit.

Receiver: In electronic systems, any device used to receive, amplify, and demodulate signals transmitted by another device.

Resistance (R): An electrical or electronic circuit component that provides resistance to current in a circuit.

Satellite Dish: This is a general term that refers to the downlink satellite antenna receiver. This can mean either the large parabolic home satellite dish (HSS), or the 18-inch digital satellite receiver (DSS).

Satellite Receiver: This is a specialized converter box that amplifies, translates and decodes the incoming video signals from the satellite dish (DSS or HSS).

Signal Strength: The intensity of a signal being transmitted from a broadcast point, amplifier, or satellite. The strength of the signal can be measured by an RF meter.

Signal-To-Noise Ratio: The ratio of the desirable portion of a signal to unwanted portion of a signal, usually expressed in decibels (dBs). The higher the number, the better.

SLM: This is the abbreviation for Signal Level Meter—an instrument used to measure TV signals.

Splitters: A device that is used to divide or split the signal from an incoming cable line. It is usually located in your attic. Splitters are available with two to four ports, which send the incoming cable signal to up to four rooms.

Tap: The point at which the cable company's distributed signal is coupled to a subscriber.

Transformers: Sometimes called a "Matching Transformer", a cylinder shaped device that has two wires coming out of the top and a threaded connector for an F connector on the bottom. The transformer enables any TV that is not cable-ready (doesn't have a 75-ohm coaxial cable connector for the antenna input) to receive a cable signal. It converts the 75-ohm coaxial cable impedance signal into a 300-ohm impedance signal to match the 300-ohm terminals of the antenna input on the TV.

Transponder A radio transmitter-receiver, usually located in a satellite, which automatically transmits a signal on the reception of a proper input signal. In a satellite, the uplink signal is received by the transponder which retransmits the signal on the downlink frequency.

Twin Lead: TV signal leadin cable that consists of two wires separated a certain distance by a plastic spine. Twin lead has an impedance of 300 ohms.

UHF: This is the abbreviation for Ultra High Frequency. The frequency range between 470 - 890 MHz that consist of channels 14 to 83.

Uplink: This refers to the signals that the various TV networks beam up to their satellites in orbit. The uplinks are then converted to downlink signals beamed back down to Earth.

VHF: This is the abbreviation for Very High Frequency. The frequency range between 34 - 216 MHz that consist of channels 2 to 13.

Wireless Cable TV: This TV video system delivers its programming via microwave signals instead of a large network of cables. The subscribers home is installed with a small microwave receiving dish about the size of a newspaper on top of his house.

Index

☞ **Dear Reader:** *We'd like your views on the books we publish.*

PROMPT® Publications, an imprint of Howard W. Sams & Company, is dedicated to bringing you timely and authoritative documentation and information you can use. You can help us in our continuing effort to meet you information needs. Please take a few moments to answer the questions below. Your answers will help us serve you better in the future.

1. What is the title of the book you purchased?_____

2. Where do you usually buy books? _____

3. Where did you buy this book? _____

4. What did you like most about the book?_____

5. What did you like least? _____

6. Is there any other information you'd like included? _____

7. In what subject areas would you like us to publish more books? (Please check the boxes next to your fields of interest.)

❏ Audio Equipment Repair ❏ Home Appliance Repair

❏ Camcorder Repair ❏ Mobile Communications

❏ Computer Hardware ❏ Security Systems

❏ Electronic Concepts Theory ❏ Sound System Installation

❏ Electronic Projects/Hobbies ❏ TV Repair

❏ Electronic Reference ❏ VCR Repair

8. Are there other subjects that you'd like to see books about? _____
_____

9. Comments _____
_____
_____

Name _____
Address _____
City _____ State/ZIP _____
Online Address _____
Would you like a *FREE* PROMPT® Publications catalog? ❏Yes ❏No
Thank you for helping us make our books better for all of our readers. Please drop this postage-paid card into the nearest mailbox.

For more information about PROMPT® Publications,
see your authorized Howard Sams distributor, or call 1-800-428-7267
for the name of your nearest PROMPT® Publications distributor.

An imprint of
Howard W. Sams & Company
A Bell Atlantic Company
2647 Waterfront Parkway, East Dr.
Suite 300
Indianapolis, IN 46214-2041

BUSINESS REPLY MAIL

FIRST CLASS MAIL   PERMIT NO. 1317   INDIANAPOLIS IN

POSTAGE WILL BE PAID BY ADDRESSEE

PROMPT PUBLICATIONS
AN IMPRINT OF HOWARD W SAMS & CO
2647 WATERFRONT PARKWAY EAST DR
SUITE 300
INDIANAPOLIS IN 46209-1418